F. C Kirkwood

A List of the Birds of Maryland

F. C Kirkwood

A List of the Birds of Maryland

ISBN/EAN: 9783744714327

Printed in Europe, USA, Canada, Australia, Japan

Cover: Foto ©berggeist007 / pixelio.de

More available books at **www.hansebooks.com**

A LIST OF THE BIRDS OF MARYLAND.

By F. C. KIRKWOOD.

The following list of the Birds of Maryland is the result of a promise made to the late Prof. Geo. L. Smith shortly before his untimely death. It includes all our regular birds and is specially intended to give the time of their arrival and departure and also their nesting periods. It also includes such stragglers as I have been able to find record of, and such as no doubt occur, but have not been recorded within our limits, our state being very deficient in ornithological lists.

It is compiled from the author's field work from January 1, 1881, to date, combined with which is that of the following gentlemen who kindly gave me the use of their collections and note books, and to whom I now extend thanks :

Wm. H. Fisher, Baltimore City.
Arthur Resler,   "   "
W. N. Wholey,   "   "
A. T. Hoen,   "   "
Geo. H. Gray,   "   "
P. T. Blogg,   "   "
J. Hall Pleasants, Jr., Baltimore City.
J. E. Tylor, Easton, Md.
H. W. Stabler, Jr., Sandy Springs, Md.
The late Edgar A. Small, Hagerstown, Md.
Robt. Shriver, Cumberland, Md.

I have also to accord my thanks to others mentioned in the text and to Mr. Robt. Ridgway, of Washington, and especially to Mr. Chas. W. Richmond, of Washington, who has kindly given his time to an ample review of the manuscript and added many important notes, giving items from the District of Columbia and surrounding country.

The large majority of my own observations during the last six years have been made in Dulaney's Valley, Baltimore County, and during the last four years I have spent two weeks each spring observing our nesting birds. In 1892 and 1893, on the waters between Chestertown and Eastern Bay ; in 1894, on Sinepuxent and Chincoteague Bays, between the Delaware and Virginia lines, and in 1895, at Vale Summit on Dan's Mountain, in Allegany County. Flying visits have been made by myself and others to various other points within the state.

I have omitted the word "shot" in a great many cases, but unless it is expressly stated that the bird was only seen, the specimen has been secured; this of course not applying to such birds as are easily identified on the wing.

In compiling the number of a "set" of eggs, the majority of notes are my own; these include all nests with a completed complement of eggs, whether collected or not, and also such nests as held young birds from which none had flown. In some of our commoner species the number is quite large, and shows the variation in a "set" much better than the usual "four or five, sometimes three or six."

I have omitted mention of "second nestings;" a good many of our birds nest twice and some three times; but except in one or two cases I could not identify the same pair of birds, and as nests in all stages, from "started to build" to "birds ready to fly," can be found any day in the height of the season for most of our commoner species, I only give extreme nesting dates, preferring such as record eggs where possible.

A list of the works and publications to which I have made reference is appended. The nomenclature followed, as also the number given, is that of the American Ornithologists Union.

The index kindly prepared by Mr. W. H. Fisher includes all our local names, each bird being referred to under its A. O. U. number.

*   *   *   *   *   *

Maryland is situated between the parallels of 37° 53' and 39° 44' northern latitude and the meridians of 75° 04' and 79° 33' western longitude. Mason and Dixon's line separates it from

Pennsylvania and Delaware, and it is separated from Virginia by a line drawn from the Atlantic Ocean to the western bank of the Potomac River and low water line on the Virginia shore, this being the southern border of Maryland from the source down. From the source of the Potomac a line runs north to Mason and Dixon's line. The gross area of the state is 12,210 square miles: 9680 land; 1203, the Maryland part of the Chesapeake; 93, Assateague Bay; and 1054, smaller estuaries and rivers.

The three leading topographical regions of the eastern portion of the United States, *viz.*, the Coastal Plain, the Piedmont Plateau, and the Apalachian region are all typically represented within the limits of the state. The Coastal Plain, or tide-water Maryland, forms the eastern portion, lying south of a line drawn from Wilmington to Washington, through Baltimore. This is closely outlined by the track of the Pennsylvania Railroad; it covers about 5000 square miles. The Eastern Shore, except in the extreme north, does not reach at any point 100 feet in elevation, while most of it is below 25 feet. It is deeply cut up by tide-water rivers and bays. This also describes the tract known as the "Necks" between Baltimore and Havre de Grace. The temperature is much modified by the surrounding water, the southern portion having a mean annual temperature of 58°; but the greater part lies between 56° and 54°; the northern part averaging 52°. Southern Maryland or the Western Shore is different, considerable of it reaches 100 feet, and in places as much as 180 feet. The mean annual temperature seldom exceeds that of Baltimore, which is 55.6°, by more than 2°.

The Piedmont Plateau, or Central Maryland, extends from the Coastal Plain to the Catoctin Mountain, and has an area of about 2500 square miles; it is broken by low, undulating hills, which gradually increase in elevation to the westward. Along the eastern margin, heights exceeding 400 feet are frequently reached, and at Catonsville 525 feet, while at Parr's Ridge, in Carroll County, it rises above 850 feet. At Frederick City, the elevation of the valley is about 250 feet above tide. The mean annual temperature ranges from 50° to 55°.

The Apalachian region, or Western Maryland, consists of a series of parallel mountain ridges with deep valleys between them, cut at nearly right angles by the Potomac River. The Catoctin Ridge reaches 1800 feet, and the Blue Ridge, at Qui-rauk, 2400; while at Middletown and Hagerstown the elevation is 500. The mountains proper begin at North Mountain, and reach in a number of cases 3000 feet and over; while the valleys near the Potomac have an elevation of 500 feet rapidly ascending; the river at Cumberland being about 600 feet. The mean average temperature ranges from 50° to 53°.

In the above mentioned areas, the Coastal Plain and the Piedmont Plateau are ornithologically within what is usually considered as the Carolina faunal area, as shown by the nesting of the Cardinal, Carolina Wren, Tufted Tit, etc. It will be found to be fairly divided into two sub-faunal regions. In the lower, or Coastal Plain, will be found the Mockingbird, Fish Hawk and Fish Crow, breeding in numbers. In the southern part of Maryland will be found traces of a still more southern faunal area, the Louisiana, distinguished by the presence of the Brown-headed Nuthatch.

The eastern part of the Apalachian region is also included in the Carolina faunal area. The Alleghanian fauna covers the mountain region from North Mountain westward, it also appears an the higher parts of the Blue Ridge. This faunal region is the breeding range of the Chestnut-sided Warbler, Rose-breasted Grosbeak, Wilson's Thrush, etc. As this section has never been fully investigated by ornithologists, there remains the possibility, as suggested by Mr. C. W. Richmond, that the Snowbird (*Junco hyemalis*) may yet be found breeding in some of the hemlock tracts still standing; this would give a tinge of a more northern faunal area, the Canadian.

A circle drawn round Baltimore City, with its centre at Baltimore and Charles streets, and a 15-mile radius, will take in over one-half of Baltimore County, and parts of Anne

Arundel and Howard Counties, a small portion of the Chesapeake Bay, and Patapsco, Back, Middle and Gunpowder Rivers. In this circle a great diversity of country is found. The part of Anne Arundel included is nearly level and low lying, as is also the "Necks" of Baltimore County. On these necks and also along as far as Havre-de-Grace, are located the famous ducking clubs of the Chesapeake. The water is brackish, or salty, rising and falling with the tide. The timber is principally short leafed pine. The other part of Baltimore and Howard Counties is more or less hilly, rising in places to over 700 feet. The water courses are swift running streams in more or less deep cuts, some even in rocky gorges. These steep banks and most of the uneven land is heavily wooded with oaks, chestnut, beech, etc. (pines only appearing in small isolated patches), while the level ground is under a high state of cultivation. All observations to which a locality is not given are within this circle; in special cases within this circle, and in all outside it, the exact location is given.

The section usually worked by Messrs. A. T. Hoen, W. N. Wholey, and myself, extends from Waverly, in the northern part of Baltimore City, to Towson, and so through Dulaney's Valley, across the lower half of Long Green Valley to the Harford pike and back to Baltimore; Mr. J. Hall Pleasants working within it around Towson.

Waverly has an elevation of 200 feet and Towson 500. The lowest point in Dulaney's Valley is the level of Loch Raven, 170 feet. The ridge between Dulaney's and Long Green Valleys reaches 560 feet. Long Green Valley is all above 250 feet. The Gunpowder River crosses this area in a zigzag course, at nearly right angles, its banks, except in Dulaney's Valley, being precipitous. Four miles of it constitutes Loch Raven.

Mr. Wm. H. Fisher's usual basis of observation is from Mount Washington to Lutherville, including Lake Roland on Jones Falls (230 feet above tide), and the Green Spring Valley, which, starting at 260 feet, near Sherwood, gradually ascends as it goes westward. The country north and south of this valley is very uneven and high (560 feet in some places).

Messrs. G. H. Gray and P. T. Blogg, working to the west of the city, have quite a different country, most of their observations being made in the rocky, heavily timbered gorge of Gwynn's Falls, between Calverton and Powhattan. The bed of the falls at Calverton has an elevation of 80 feet, and at Powhatan dam of 320 feet, the country on both sides ranging from 300 to 500 feet.

Mr. A. Resler takes another field, collecting a good deal on Patapsco marsh, just south of the city, also "down the necks," and near Botterill Post-Office, in Howard County, 14 miles from Baltimore.

Order PYGOPODES—Diving Birds.

Family Podicipidæ—Grebes.

Colymbus holbœllii (2).   Holbœll's Grebe.

"Not uncommon on the Potomac in winter" (A. C., 110), this species has been taken in Lancaster County, Pa., and at other points in Pennsylvania and New Jersey (Birds E. Pa. and N. J., 38), so I presume it occurs all over tide water Maryland. On February 25, 1894, I watched one for about an hour at Lake Roland where it was swimming around in a hole in the ice, which otherwise covered the lake. Later, when we threw stones at it, it would neither fly nor dive, so we left it.

Colymbus auritus (3).   Horned Grebe.

Fairly common in tide-water Maryland during migrations, a number winter with us. Noted from October 7 ('76, Resler) at Back River, to April 23 ('90, Resler), at the same place. April 25, at Washington, D. C. (Richmond). Inland, a young male, shot at New Market, Carroll County, by Dr. H. H. Hopkins, was presented in the flesh on February 21, '81, to the Maryland Academy of Sciences. A pair, male and female, were taken at Hagerstown on April 16, '83 (Small), and Mr. Robert Shriver has secured specimens at Cumberland.

Podilymbus podiceps (6).   Pied-billed Grebe.

Common from September 1 ('93) to March 31 ('88, Resler), a few stay with us to breed, but, as far as I know, the
nest has not been found in Maryland. Specimens have been
taken on April 28 ('93) and July 31 ('75, Resler), at Back
River. At Hagerstown, in June ('80, Small), and near
Westminster, early in September ('80, Fisher) nine, possibly a
family, were taken in one day from a mill pond. At Washington, common from August 25 to May (Richmond).

Family URINATORIDÆ—Loons.

Urinator imber (7).   Loon.

Fairly common during winter on ocean front, Chesapeake
Bay, and larger waters of Maryland. In New Jersey it is
given as arriving October 3 (Birds E. Pa. and N. J., 39), but
I have only spring dates ranging from March 9 ('91, Fisher), at
Legoe's Point, to June 17 ('93, J. F. Hargreaves), when a very
noisy pair were on the Gunpowder River, near the Pennsylvania
Railroad bridge. Mr. W. S. Walker, of Chestertown, writes
me that it is "one of the last birds to leave Chester River."
At Washington, from September to April 25 (Richmond).
Audubon says (vii, 284): "the Loon breeds in various parts of
the United States from Maryland to Maine. I have ascertained
that it nestles in the former of these states on the Susquehanna
River." Not known to nest here now.

Urinator lumme (11).   Red-throated Loon.

Occasionally taken on our waters during winter. On February
16, 1878, the late A. Wolle presented the Maryland Academy
of Sciences with shells taken from the stomach of one of these
birds, presumably captured near Baltimore.

"Not uncommon on the (Potomac) River during the winter
months" (A. C., 110). "In the spring of 1882 one was caught in
a gill net in the Potomac River, a few miles below Washington,
and is now in the possession of Mr. O. N. Bryan, of Marshall

Hall, Md." (H. M. Smith and W. Palmer, Auk, v, 147).   One
shot on Chester River was sent to the Smithsonian for idendi-
fication by Hiram Brown, Pomona, Md. (Smith. Report, 1885,
192), "and I also learn from Dr. A. K. Fisher of another, captured on the Potomac, near Fort Washington, October 20, 1889"
(Birds Vas., 41).

Family ALCIDÆ—Auks, Murres, and Puffins.

Cepphus grylle (27).   Black Guillemot.

South in winter to New Jersey (Manual, 16).   Audubon
says (vii, 273), "during severe winters, I have seen the Black
Guillemot playing over the water as far south as the shores of
Maryland.   Such excursions, however, are of rare occurence."

Alca torda (32).   Razor-billed Auk.

"Winter visitant on the New Jersey coast; one specimen was
secured by Dr. W. L. Abbott as far south as Cape May"
(Birds E. Pa. and N. J., 41).   "Capt. Chas. H. Crumb informs me that three have been taken near Cobb's Island, two
in 1884 and one in 1887, one of which is now in my possession"
(Birds Vas., 4), and several more were taken during the winter
of 1892–93 (Capt. Crumb in letter to Wm. H. Fisher).

Alle alle (34).   Dovekie.

"Regular winter visitant along the New Jersey coast, varying
in abundance from year to year" (Birds E. Pa. and N. J., 41).
On December 9, 1877, one was caught alive on the sea beach
near Ocean City, Md., and sent to the Maryland Academy of
Sciences by Mr. Robert Henry, of Berlin, Md.   Capt. Crumb
has taken two at Cobb's Island.

Order LONGIPENNES—LONG-WINGED SWIMMERS.

Family STERCORARIIDÆ—Jaegers and Skuas.

Stercorarius pomarinus (36).   Pomarine Jaeger.

Winters on the Atlantic coast from Long Island southward.

Stercorarius parasiticus (37).   Parasitic Jaeger.

Winters along the Atlantic coast from the middle states southward.

Stercorarius longicaudus (38).   Long-tailed Jaeger.

Migrates south along the Atlantic coast to the Gulf of Mexico and West Indies.
No doubt all three of the Jaegers occur off our coast, but we have no record of them.

Family LARIDÆ—Gulls and Terns.

Rissa tridactyla (40).   Kittiwake.

"Very rare winter visitant on the New Jersey coast" (Birds E. Pa. and N. J., 42). "About ten years ago the late Henry B. Graves, of Berk's County, mounted a young Kittiwake which had been captured near Lancaster City, in midwinter," and "Dr. A. C. Treichler, of Elizabethtown, mentions the species as a straggler in Lancaster County, Pa." (Birds Pa., 17). "Captain Crumb reports this species as a rare and irregular winter visitant at Cobb's Island, but he has never taken a specimen" (Birds Vas., 41).

Larus leucopterus (43).   Iceland Gull (?).

On November 23, 1893, I saw a pure white gull in the inner harbor of Baltimore City. It came within fifty feet of me at times, as I watched it for fully half an hour. In reply to a description of this bird, which I sent to Mr. Robert Ridgway, he writes : "The gull which you think may be the young of *L. leucopterus*, is undoubtedly what American ornithologists here consider and describe as the young of that species. There is a question, however, whether it is not in reality the young of *L. kumlieni*. There are no present means of settling the question, there being no specimens of undoubted young of *L. leucopterus* in any American collection so far as I know."

Larus marinus (47).   Great Black-backed Gull.

"Rare winter visitant along the New Jersey coast, a few, however, probably occur every year" (Birds E. Pa. and N. J., 43). "Has been taken (including the adult) at Cobb's Island, where on the authority of Capt. Crumb it is not common, though seen every winter" (Birds Vas., 41) ; "south coastwise in winter to Florida" (Key, p. 743). On January 27, 1895, at Holly Point, I saw one of these birds circling over the mouth of Gunpowder River, but it kept considerably out of range.

Larus argentatus smithsonianus (51a).   American Herring Gull.

In Baltimore harbor this species is a common winter resident, common over the Basin at Light and Pratt streets, where they live on the refuse. During the latter part of September I could not find any, but quite a number were at hand on October 5, ('94). On May 1, ('95), a number were observed, and on May 6, ('95), two were in Canton Hollow. They also occur on all our waters, but not in numbers as they appear in our harbor. "Common in winter at Washington" (Richmond).

Larus delawarensis (54).   Ring-billed Gull.

At Washington "seen over the river during the winter months, more frequently than the Herring Gull" (A. C., 108). "Winters abundantly on the coast of the Middle States, I saw it continually during two winters over the harbor of Baltimore where it flies among the shipping, with Bonaparte's Gulls and several kinds of terns" (Birds N. W., 638).

However the above may have been when Dr. Coues was at Fort McHenry, I have been unable to substantiate the statement, a stray tern being of very unusual occurrence above the Fort, while Bonaparte's Gull, as a rule, keeps below it. My only spring notes are of two near Holly Point over Gunpowder River on March 25 ('94), and a few in the lower harbor on March

29 ('95), but in fall they are noted near Fort McHenry from September 26 ('94) to November 6 ('94).

Mr. Ridgway writes me: "As to *Larus delawarensis* wintering near Baltimore, it may occasionally do so, but not in any considerable numbers. Here, on the Potomac, it is not considered a winter resident, though it migrates northward very early in spring," and Mr. C. W. Richmond writes: "Not a winter visitant at Washington, it was common during March until the 30th, 1890. Noted from February to April 5, and again in October and November."

Larus atricilla (58).   Laughing Gull.

Common migrant from April 28 ('94, Wholey) to May 22 ('95), and from September 29 ('94) to October 12 ('94). A few may spend the summer with us, two pairs being noted at Patapsco Marsh on July 3 ('93, Blogg).

"One of the most abundant gulls at Cobb's Island, where numbers commence to breed about the 20th of June" (Birds Vas., 41). I have been told they breed at Chincoteague Island, and from the fact that I saw a few flying north in the morning and south in the evening over Chincoteague Bay, in Maryland, I presume they do, though a search of ten days (June 5 to 14, '94) failed to locate a breeding site on the ocean front of Maryland, and I also failed during two seasons ('92 and '93) to locate a breeding site on the Chesapeake, though frequently told of them by the fishermen, and frequently seeing paired birds.

Larus philadelphia (60).   Bonaparte's Gull.

A tolerably common migrant, noted from March 24 ('94, Fisher) to May 17 ('93), and from October 5 ('94) to November 9 ('92, Resler); a few may also winter with us (Birds N.W., 638). While this species may come up the harbor with the Herring Gulls, I do not think it ever comes beyond the broad water at Broadway Ferry. At Washington they are given as "comparatively common in August and September" (A. C., 108).

Gelochelidon nilotica (63). Gull-billed Tern.

"Rare visitant along the New Jersey coast, where it is reported to have bred formerly ('As late as 1886, according to Mr. H. G. Parker' [O. and O., 1886, p. 138])." (Birds E. Pa. and N. J., 44.)

On May 19, 20, 21, 1892, Mr. W. H. Fisher noted this species at Cobb's Island, Va., where, Mr. H. B. Bailey says, "a few pairs were seen, but they had not commenced to breed during my visit, May 25 to 29, 1875. They nest here sparingly, however, as I had a set of their eggs sent me which were laid the last of June" (Auk, i, 24–28), and, May 14 to 28, 1894, they "seem to be rapidly diminishing in numbers, being far less numerous than I observed them on two previous trips in 1891 and 1892, when I was collecting in vicinity of Smith's Island, Va." (E. J. Brown, Auk, xi, 259). Casual at Washington (Richmond).

Sterna tschegrava (64). Caspian Tern.

Rare migrant along the Atlantic coast.

On July 4, 1880, Mr. Robert Ridgway found two nests with two eggs in each, they were at opposite ends of Cobb's Island, and about ten miles apart (B. N. O. C., v, 221–22–23). "Capt. Crumb has found three sets of eggs in July" (Birds Vas., p. 42).

"Hon. J. J. Libhart, in his ornithological report, published in the history of Lancaster County, records the capture of two of these birds on the Susquehanna at Marietta, Pa., on September 21, 1847" (Birds Pa., 19).

Sterna maxima (65). Royal Tern.

"They have always been found breeding on a small sand bar off Cobb's Island, but it was washed away during the winter of 1874–75, and although the birds were flying about (May 25–29, '75), they had not chosen any spot on which to breed; but they undoubtedly did so later" (H. B. Bailey, Auk, i, 24–28).

On July 4, 1880, Mr. Robert Ridgway saw an immense colony there (B. N. O. C., v, 221–22–23), and "Captain Crumb has found eggs in the latter part of June" (Birds Vas., 42)." "Rare straggler on the New Jersey coast during summer" (Birds E. Pa. and N. J., 45).

Sterna forsteri (69).   Forster's Tern.

"It is the commonest tern in winter and during the migrations in the harbor of Baltimore" (Birds N. W., 679). On March 4, 1893, during a severe snow storm, I saw several terns flying over the channel off Sparrow's Point, which I believe were of this species, but they did not come close enough to be positively identified.

They have been taken near Washington, D. C., and have been found quite abundant over the lower Potomac by Mr. P. L. Jouy (Field and Forrest, vii, 29).

On the evening of June 5, 1894, at North Beach, about ten miles south of Ocean City, Md., I saw some boys who had gathered about 200 "Striker" eggs, and were proceeding to cook same for their supper; they had also shot a number of birds which, on examination, I found to be all Forster's Terns. Next day, June 6, 1894, I visited a marshy island with probably 2000 terns over it, and as far as I could observe all were *forsteri*. Here I noted 12 nests with 3 eggs, 19 with 2 and 41 with 1; how many without eggs I am unable to say.

On June 7 I visited two other much smaller "tumps" about three miles distant; about 200 birds here, on one there were 7 nests with 1 egg each, and on the other 7 with 1, and 1 with 2; this was evidently where the boys had been, as innumerable nests were empty. On June 10 I again visited all three of these "tumps" and on the larger one noted 7 nests with 3 eggs, 7 with 2, and 19 with 1, the number of unoccupied nests being greater than before. On the smaller "tumps" I saw respectively 1 of 2, and 6 of 1; and 3 of 2 and 4 of 1; and more unoccupied nests. Here the birds were much fewer than before, while at the largest "tump" there appeared to be more than on the 6th. These nesting places had been visited by other parties between my visits.

Sterna hirundo (70).   Common Tern.

"In North America chiefly confined to the Eastern Provinces, breeding variously throughout its range" (A. O. U.). Apparently migratory, three specimens were taken on May 6 ('76, Resler) out of a number seen at Patapsco marsh ; and on May 17 ('93), one was shot at the same place. On September 1 ('93, A. Wolle), four were shot out of about fifteen at Gunpowder marsh, these I examined.

Sometimes common, but an irregular migrant at Washington (Richmond).

Sterna paradisæa (71).   Arctic Tern.

Breeding from Massachusettes northward, this species comes south in winter along the Atlantic coast at least as far as Virginia.

Sterna dougalli (72).   Roseate Tern.

This southern species, going regularly north in summer to Maine, "doubtless breeds" at Cobb's Island (H. B. Bailey, Auk, i, 24–28), and is a "rare straggler on the New Jersey coast during summer ; formerly it is reported to have bred in considerable numbers" (Birds E. P. and N. J., 47).

Sterna antillarum (74).   Least Tern.

A summer resident in restricted localities, this species is usually seen during migrations.

On June 13, ('94), I visited a nesting colony at the juncture of Miles River with Eastern Bay, where, on a small island, I found four sets of three ; twelve of two, and eight single eggs; how many other nesting hollows it is impossible to say, as they are so slight they show no signs of a nest until an egg is deposited. During the fall of 1893, this species was noted at Lake Roland on August 19 and 20, and September 3 and 4 (Fisher).

Sterna fuliginosa (75).   Sooty Tern.

North to the Carolinas and casually to New England; "there is a specimen from Baltimore in the National Museum at Washington, obtained from the late Mr. A. Wolle (Robert Ridgway).

"Dr. A. C. Treichler mentions it as a straggler in the neighborhood of Elizabethtown, Lancaster Co., Pa." (Birds Pa., 23).

Hydrochelidon nigra surinamensis (77).   Black Tern.

While this species is locally common over the whole of North America, I can find but few records for the vicinity of Maryland. At Washington, it is given as "less numerous than the Least Tern, found at same seasons" (A.C., 109) " Not uncommon in the early fall at Cobb's Island, Va. It has been seen there in the breeding season, but is not known to breed " (Birds Vas., 43). "Transient, occurring on the New Jersey coast, but much less abundantly than formerly. Mr. Scott states that it arrived at Long Branch in 1879, about June 11, and soon became common ; although it remained all summer, it was not known to breed (B. N. O. C., '79)" (Birds E. Pa. and N. J., 48).

The Smithsonian acknowledged "Skin of short-tailed Tern, from Potomac River, Thos. Marron" (Smith. Rep. 1891, 793).

On May 17, 1893, Alex. Wolle shot one on Patapsco marsh, and on August 25, 1893, Richard Cantter shot another near Upper Marlborough ; both of these I examined in the flesh.

Family RYNCHOPIDÆ—Skimmers.

Rynchops nigra (80).   Black Skimmer.

This strictly maritime bird of our southern coast goes regularly north to New Jersey, where it is recorded from June 10 until September 25 (Birds E. Pa. and N. J., 48).   On June 19, 1880, a specimen was presented to the Maryland Academy of Sciences, presumably taken near Baltimore.

"Individuals were once seen by ourselves on the Potomac,

some distance below Washington, September 8, 1858 " (A. C., 109).

At Cobb's Island, Va., from May 25 to 29, 1875, they were found in flocks of 20 or 30, as they do not breed until the last of June (H. B. Bailey, Auk, i, 28), and just before dusk on May 20, 1891, four were seen there (Fisher).

"Common at Smith's Island, Va., where we got three on May 15, 1894, but they had not begun to breed by the 26th when we left " (C. W. Richmond).

Order TUBINARES—Tube-nosed Swimmers.

Family PROCELLARIIDÆ—Shearwaters and Petrels.

Puffinus major (89).   Greater Shearwater.

"Atlantic coast generally. A rare straggler to the New Jersey coast " (Birds E. Pa. and N. J., 49).

Puffinus auduboni (92).   Audubon's Shearwater.

Atlantic coast from New Jersey southward, breeds in the Bermudas and Bahamas (A. O. U.).

Puffinus stricklandi (94).   Sooty Shearwater.

North Atlantic, south to the Carolinas, breeding far north. No doubt all three occur off our ocean front, but as there are no observers there, their presence with that of other of our ocean birds has not been recorded.

Oceanodroma leucorhoa (106).   Leach's Petrel.

On June 11, 1895, while fishing on Little Gull Bank, about three miles out from Ocean City, a pair of these birds came and circled round our boat for a few minutes. I have also noted them further out, off our Maryland coast, on several occasions during the month of August, when they followed the vessel generally in company with Wilson's Petrel.

Occasionally heavy easterly gales drive them inland; at Washington, D. C., "one of several shot in August, 1842, is now in the National Museum, also another taken near the Navy Yard Bridge, on June 5, 1891, by Wm. Bayley. Wm. Palmer has two; one was shot from a bunch of three or four on August 29, 1893, the other was captured alive in a house on Capitol Hill a few days later" (Richmond).

Oceanites oceanicus (109). Wilson's Petrel.

I have been greatly interested in watching these birds as they flew round the vessel, on several sea trips I have made. Under date of August 20, 1884, I find the following: "Cleared the capes last night at 11.30 P. M., and this morning we have the petrels, which stay with us all day, the Maryland shore being in sight. One is partly albino, having a white breast, belly and back, separated from the usual white by a sooty line."

The only inland record I can find is "one taken many years ago and presented to the Smithsonian" (A. C., 110). "This was shot on the Potomac River about 1859" (Richmond).

Order STEGANOPODES—TOTIPALMATE SWIMMERS.

Family SULIDÆ—Gannets.

Sula bassana (117). Gannet

Occasionally taken on the Chesapeake, three specimens so recorded have been presented to the Maryland Academy of Sciences, where two of them are at present, the other, now in the Johns Hopkins University, taken many years ago at Chestertown, was presented by Mr. Colin Stam(sen?) through Mr. J. J. Thomsen. On April 21, 1894, one was taken in Rock Hall Cove and forwarded alive to the Academy by Dr. A. P. Sharp. Mr. John Murdock presented a mounted specimen on June 6, 1892, referring to which he writes me: "The bird I presented to the Maryland Academy of Sciences was killed a short time previously in the lower part of the Chesapeake Bay, near Mobjack Bay; I have never seen them above that point; generally they go in pairs."

"I found the remains of one on Smith's Island, Va., in May, 1894. It had been dead about a month, possibly more" (Richmond).

Family ANHINGIDÆ—Darters.

Anhinga anhinga (118).   Anhinga.

In the old collection of the Maryland Academy of Sciences was a mounted specimen of the Anhinga, which Prof. Uhler says, came from the Pocomoke River, but owing to the vicissitudes through which the Academy has passed, I have been unable to find the record of its acquisition.

Family PHALACROCORACIDÆ—Cormorants.

Phalacrocorax carbo (119).   Cormorant.

Audubon says, "it is rarely seen further south than the extreme limits of Maryland, but from Chesapeake Bay eastward it becomes more plentiful" (vii, 418). A specimen may occasionally visit us with the following species, as it goes casually south as far as the Carolinas (A. O. U.).

Phalacrocorax dilophus (120).   Double-crested Cormorant.

Regular, but not a common winter visitant near Baltimore. Further down the Chesapeake, and on Chincoteague and Sinepuxent Bays, it is more numerous. Under date of April 13, 1893, Mr. Wm. S. Walker, of Chestertown, writes me: "The only Cormorant I ever had in hand, I killed some five or six years ago at Hail Creek, at the mouth of Chester River. I have since that time seen one or more of the birds sitting on buoys in the bay between here and Baltimore." The specimen mentioned was in the old collection of the Maryland Academy of Sciences.

At Ocean City single birds flew northward on June 8 and 10, '94.

"One was detected in the District of Columbia many years ago" (A. C., 108). Mr. Geo. W. Duvall sent a specimen from Annapolis to the Smithsonian (Smith. Rep., '72, 57).

Family PELECANIDÆ—Pelicans.

Pelecanus erythrorhynchos (125). American White Pelican.

Rare straggler. "There appear to be three well authenticated instances of the capture of this bird in our vicinity. 1. Near Alexandria, Va., April, 1864, by C. Drexler, and presented to the Smithsonian. 2. Opposite Washington, on the Virginia bank of the Potomac, fall of 1864; shot by John Ferguson, and seen and identified by several persons who have communicated the fact to us. 3. Near Alexandria, Va., October, 1878, killed by John Haxhurst, and seen by a gentleman connected with the National Museum" (H. M. Smith and Wm. Palmer, Auk, v, 147). "A stray Pelican at Oakland, Md., by Sportsman," is recorded (Auk, iv, 345).

Pelecanus fuscus (126). Brown Pelican.

Exclusively maritime, "from Tropical America to the Carolinas" (A. O. U.). Captain Crumb noted them at Cobb's Island, in the fall of 1881 (Birds Vas., 44), and "Turnbull records one specimen shot off Sandy Hook in 1837, in summer" (Birds E. Pa. and N. J., 52). In the old collection of the Academy was a specimen from the lower Potomac (Uhler).

Order ANSERES—LAMELLIROSTRAL SWIMMERS.

Family ANATIDÆ—Ducks, Geese, and Swans.

Merganser americanus (129). American Merganser.

This, the largest of the "fishermen," is a winter resident, from Sept. 29, ('94), to March 29, ('93 Resler); common on our larger waters.

On March 24 ('95) I watched a bunch of eight for quite a while on the Gunpowder Falls; they were diving in the swift running stream, just above the dead water of Loch Raven. Mr. Shriver says, "something of a rarity at Cumberland, but I have seen a number of them years ago; none lately."

"In the ornothological report of the late Judge Libhart, published about twelve or fifteen years ago in the history of Lancaster county, the Goosander, also the Red-breasted and Hooded Mergansers are all mentioned as breeding in Lancaster County." (Birds Pa. 32).

Merganser serrator (130).   Red-breasted Merganser.

From the number brought to market, this is the most common "fisherman" shot on the Chesapeake and its tributaries during the gunning season.   Possibly it also stays here to breed, as Mr. W. N. Wholey shot a female on July 8, 1892, at Egging Beach Island, near Ocean City.   It was in good condition and not crippled in any way, but the breast feathers were so few and so worn that there can be no doubt of its having nested. Two others flushed at the shot but were not secured.   Loch Raven (Chas. E. Dukehart).

Lophodytes cucullatus (131).   Hooded Merganser.

Common on the arms of the Chesapeake during gunning season ; this species is also to be found generally dispersed over the state, even on small waters.   Several flocks were seen on the Potomac, about two miles below Knoxville, on November 5 ('93, Fisher).   Loch Raven (Dukehart).

It probably breeds, for on June 7 ('94) I noted one at Whittington's Point, near Ocean City, and Dr. Warren says: "I have an adult female, taken June 23, 1890, in Chester Co., Pa., where this Merganser is seldom seen in summer" (Birds Pa., 345).

Anas boschas (132).   Mallard.

Common winter resident, from October 7 ('93) to April 14 ('94, Fisher).   Mr. W. S. Walker, of Chestertown, says: "the first ducks to arrive here are the 'flat fowl,' that is those feeding on the flats of the Bay shore, they are Black Ducks, Mallards, Graybacks and Sprigtails."   A number have been shot on Loch Raven (Dukehart), and on November 4 ('93, Fisher) they were very numerous on the Potomac between Knoxville and Brunswick.   Under "County news," on April 13, 1894, the *Sun* paper says: "Gunners in Alleghany county are having a fine time, as the streams are covered with Mallards." On April 5, 1895, a pair were shot at Cumberland (Zacharia Laney).

Anas obscura (133).   Black Duck.

Common in tidewater Maryland during the gunning season, quite a number remain during summer and breed. Numerous in the vicinity of Baltimore from August 28 ('93, Fisher), to May 6 ('93, Gray). I observed this species in 1894, at Ocean City, as follows: On June 5 a pair; on the 6th a pair; on the 7th 3, 2, 3, 1, 3, and also a pair with small young on Chincoteague Bay about two miles from shore, where it was so rough I could not count the young. On the 10th my man at last succeeded in finding a nest, but it only held the two half shells of an egg, one inside the other.

At Loch Raven, on April 7, ('95), I watched 20 birds for some time, they were feeding in shallow water like tame ducks, and while they kept in a close bunch, were mated, each pair distinctly keeping together.

Anas strepera (135).   Gadwall.

While a few no doubt winter with us, the species seems to be fairly numerous in November, and again in April; though it is rather difficult to get statistics, our market gunners classing this and the females of three or four other species as "gray ducks," and our amateurs as "trash ducks." On November 22–3, 1894, about 20 were shot at Spry's Island. At Washington, D. C., they are given as common from September to April (Richmond).

From the *Sun* I take the following: "Williamsport, Md. April 9, 1895. Wild ducks are plentiful along the Potomac, above this place. From the old Sharpless warehouse up to Big Pool, flocks of Gray Mallards abound. These, the sportsmen say, are rare, and it is unusual to see so many of them along the river."

Anas penelope (136). Widgeon.

This Old World species is given as "rare or casual along the Atlantic coast of North America" (Key, 694).

Several specimens have been taken within our state. One found in market, at Washington by Mr. C. Drexler, in the spring of 1863, shot near Alexandria, Va., is now in the National Museum (A. C., 103–4). On July 11, 1890, an adult male was found in Washington Market, New York. It "had been shipped from Baltimore and doubtless was shot on the Chesapeake Bay," and is now in the American Museum of Natural History, N. Y. (Edgar A. Mearns, Auk. viii, 204). "On the property of the Carroll's Island Club, Baltimore County, Mr. Wm. Carpenter, on February 25, 1890, killed one of this species from a bunch of Baldpates. This bird has been mounted and is at present in my care" (L. S. Foster, Auk, viii, 283).

At a stated meeting of the Maryland Academy of Sciences, held April 5, 1880, Mr. Arthur Resler referred to a specimen of the European Baldpate which he had examined at the taxidermist store of A. Wolle, where it had been sent to be mounted. It was shot on the Atlantic coast of Maryland (Minutes of Maryland Academy of Sciences, 1880, 280).

Dr. Wm. H. Poplar has a specimen in his house at Havre-de-Grace. He told me that he shot that "Red-headed Baldpate" in November, 1881, and considered it a cross between a Redhead and a Baldpate.

Anas americana (137). Baldpate.

Common during winter, this species, as with nearly all of our ducks, is most numerous during fall and spring flights.

On September 23 ('93, Fisher) several bunches were noted at Sparrows' Point, and the last Baldpate was shot at Grace's Quarter on April 8 ('86, Ducking Record).

Inland, a male was taken in October '88, at Ridgley's Dam in Dulaney's Valley (Fisher). On November 4, '93, several bunches were on the Potomac, near Brunswick (Fisher), and it has been taken at Cumberland (Shriver).

On October 3, '89, I witnessed a remarkable instance of the well-known sailing power of a dead duck, if shot in the heart with its wings spread. Fishing near Maxwell's Point, my attention was taken by several shots on Saltpeter Creek, and looking in that direction I noticed a duck coming down the wind, which was blowing fresh from a few points N. of W. As it got closer I noticed it was sailing, wings and neck at full stretch; when first seen it was well up and over the land, but gradually descending, it struck the water with a splash about one-quarter mile from our boat and near the centre of the river, or about two-and-a-half miles from where it was shot, the neck being about a mile across and Gunpowder River about three miles wide at this point.

Anas crecca (138).   European Teal.

One shot on the Potomac River, near Washington, in April, 1888, was presented to the National Museum (Auk, iii, 139). It was an "adult male, shot by Henry Marshall, of Laurel, Md." (Smith. Report, '86, 154).

Anas carolinensis (139).   Green-winged Teal.

Common during spring and fall migrations, a number stay in tidewater Maryland during mild winters. It is not so numerous as the following species, with which it arrives and departs. Mr. W. S. Walker, of Chestertown, writes me: "Among the last to leave Chester River is the Teal, or Partridge Duck, a little brown duck."

Anas discors (140).   Blue-winged Teal.

Usually seen with, or in the same places as the last mentioned, and during the same time of year. A bunch of 15 were noted in Bear Creek, on August 20 ('93, Fisher). The latest date is May 7 ('90, Resler), when one was taken on Patapsco Marsh.

On September 17, '93, I saw a bunch of 10 on Loch Raven, in Dulaney's Valley. They were standing at edge of water, and allowed me to walk up opposite them. After observing them through a field glass for some time, I flushed them with a stone and they flew up stream a short distance. Following them in full view, I again got opposite and sat down, they appeared quite tame, and although my dogs paddled round in the water near them, they did not fly until I again threw a stone among them ; evidently they had just arrived from the north and were tired out.

At Hagerstown, during April ('80, Small), and at Cumberland (Shriver).

Spatula clypeata (142).   Shoveler.

Quite a number of "Broadbills" are shot during our gunning season, but they cannot be said to be common.   I have no characteristic dates.   A pair were taken at Gunpowder River, on March 27, ('95, Resler).   Not uncommon at Washington during winter (Richmond); at Cumberland (Shriver).

Dafila acuta (143).   Pintail.

Common during fall and spring flights, I believe some stay with us during mild winters.   Mr. N. S. Bogle, of Eastern Neck Island, writes me, that "a flock of 15 Sprigtails arrived in Chester River, on September 13 ('93), these were the first ducks."   They are noted from that on to October 28 ('93). On March 4 ('94, Fisher), about 20 were in a bunch on Bird River, and several were taken on Choptank River on April 7 ('94).   Loch Raven (Dukehart).

Aix sponsa (144).   Wood Duck.

Sparingly resident, this species is numerous in spring and fall. While usually seen "down the necks," single birds or pairs are often flushed from pools on very small runs, the principal attraction apparently being oak woods.   Some years ago, in May, on Hog Creek, a female and eight young were observed (Fisher), and Mr. P. A. Bowen, writing from Aquasco, Prince George's County says, "resident, there are now in this neighborhood three birds, hatched in confinement from eggs taken from a hollow tree."

Resident but not common, at Washington (Richmond); Hagerstown (Small) ; Cumberland (Shriver).

Aythya americana (146). Redhead.

Common during winter in tidewater Maryland. They arrive earlier, but the first date I have is October 3 ('89), when an enormous number of ducks were "bedded" on Gunpowder River, below Maxwell's Point, quite a large number of them being Redheads. As late as May 2 ('95, Tylor), about 150 were on Gunpowder, near Magnolia, in five or six small bunches. Inland, they are numerously recorded. About 10 years ago 500 or 600 stopped for a few days on Lake Roland (Fisher); on March 29 ('91), several small bunches were on Loch Raven (Wholey), and a number have been taken in Dulaney's Valley (Dukehurt). On November 4 (93), a few were on the Potomac near Knoxville (Fisher), and they are given as occasional at Cumberland (Shriver).

Aythya vallisneria (147). Canvasback.

Still a common winter resident of tidewater Maryland, though most numerous during fall and spring flights.

On October 3, 1889, quite a large number where with the Redheads, Blackheads, etc., bedded on Gunpowder River, and they have been taken at Grace's Quarter from October 21 ('80) to April 4 ('86, Ducking Record), but no doubt some remain later.

Aythya marila nearctica (148). American Scaup Duck.

Common on the Chesapeake and its numerous arms during winter. On October 3, 1889, a number were with the other ducks on Gunpowder River, and on March 25, 1894, I watched a most affectionate pair, male and female, for some time at Cedar Point. It also occurs inland, about 200 were on Lake Roland October 29 ('92, Fisher). Dr. Owings, of Ellicott

City, shot one on his ice-pond in November, 1893. Loch Raven (Dukehart); Harpers Ferry, a bunch of six, November 6 ('93, Fisher); Cumberland (Shriver).

Aythya affinis (149). Lesser Scaup Duck.

Arriving and departing with the former species, this is much the more numerous, and both with the following, are generally classed together by gunners. At Grace's Quarter, "Blackheads" are recorded as being shot from October 3 ('89) to April 8 ('86) and, as with all other ducks, the "Blackheads" are most numerous during the fall and spring flights. On April 2 ('90, Richmond) one was taken near Washington. On April 5 ('95) one was taken at Cumberland (Zacharia Laney).

Aythya collaris (150). Ring-necked Duck.

Not very numerous. I presume it arrives with the other "Blackheads." All the notes I have range between March 8 ('94), when one was shot at Bush River by Mr. Melville Wilson, who called it a "Creek Blackhead," and April 4 ('91, Wholey), when one was shot in the open river below Fort McHenry.

Glaucionetta clangula americana (151). American Golden-eye.

Common in tidewater Maryland, from October to April, I have not any extreme dates. Mr. Dukehart has frequently shot them at Loch Raven and they have been taken at Hagerstown on December 28, 1879, and April 15, 1883 (Small).

Glaucionetta islandica (152). Barrow's Golden-eye.

This northern species coming south in winter to New York, has in one recorded instance straggled as far south as Maryland. "A female shot on the Potomac River, opposite Washington, Nov. 22, 1889, by C. Herbert, is now in the collection of J. D. Figgens" (C. W. Richmond, Auk, viii, 112).

Charitonetta albeola (153).   Bufflehead.

Common in winter, arriving the latter part of September, one was at Patapsco marsh on May 7 ('90, Resler).

On February 15, 1895, just after the blizzard, one came up the harbor into the upper · basin and stayed round off Bowley's wharf for a long time.   Inland, it has been taken at Sandy Springs in April (Stabler); a pair were seen near Harper's Ferry on October 10 ('93 Fisher), and they have been shot at Cumberland (Shriver).

Clangula hyemalis (154).   Old Squaw.

Common in winter on the Chesapeake ; I noted one at Fort McHenry on October 27, ('94), and they remain with us until late in April.   On March 4, 1895, there were several hundred in the mouth of Chester River on broad water, nearly all in pairs, but occasionally a few males were in a bunch by themselves. As the steamboat was running before a terrific snow squall, we got quite close before they flushed.   Often they dived and on coming up appeared to be flying before they emerged.   On the 6th it was blowing a gentle breeze, and although they were extremely numerous, we did not get close to any.

Loch Raven (Dukehart).

Camptolaimus labradorius (156).   Labrador Duck.

Now extinct, they were apparently quite numerous in Audubon's time, for he says (vi, 329): " The range of this species along the shores does not extend further southward than Chesapeake Bay, where I have seen some near the influx of the St. James River.   I have also met with several in the Baltimore market."

Somateria dresseri (160).   American Eider.

" Winters southward to Delaware" (Chapman, 117).   "In the old collection of the Maryland Academy of Sciences was a specimen from Maryland, collected at Pamunkey Neck, below Marshall Hall, by Mr. Chapman" (P. R. Uhler).

Somateria spectabilis (162). King Eider.

"South casually to New Jersey, in winter" (Manual, 110). A young bird of this species was obtained at Cobb's Island, Va., by Captain C. H. Crumb, on December 19, 1899" (Birds Vas., 48).

Oidemia americana (163). American Scoter.

Common during fall and spring flights, on ocean front and broad waters of tidewater Maryland, a few may winter. On November 4, ('94) four spent the evening in the basin, and on November 17 ('77), Major Hill presented a specimen in the flesh to the Maryland Academy of Sciences. On March 8 ('94), one was in Saltpeter Creek, and they were numerous on open water. Late in April, and early in June they are reported as being fairly numerous at Havre de Grace.

During Christmas week ('90) they were fairly common at Eastville, Northampton County, Va. (Ridgley Duvall, Jr.).

Oidemia deglandi (165). White-winged Scoter.

Fairly common on the Chesapeake during spring and fall, this species may winter with us. On September 12 ('94), four were noted in Chester River by Mr. E. Speddin of the tugboat Chicago, and on November 6 ('94), a pair were shot down the Necks, which I saw in market the next day.

On March 6 ('93), I saw three in a bunch off Love Point lighthouse, at the mouth of Chester River, they did not flush until the steamboat was close to them, in marked contrast to the South Southerlies; on May 12 ('94, Fisher), 3 were at the mouth of the Patapsco.

Oidemia perspicillata (166). Surf Scoter.

Arriving and departing with the others, I can give no characteristic dates. The three being as a rule classed as "tar-pots" or "bay muscoveys" and not being considered good for anything are seldom shot.

"Common winter resident off the New Jersey coast from October to late in April. Apparently the most abundant species of Scoter" (Birds E. P. and N. J., 59).

Erismatura rubida (167). Ruddy Duck.

Common during spring and fall. I have no winter dates, though possibly it remains with us. A bunch of 4 or 5 was seen in Baltimore harbor close to the Pennsylvania Railroad pier, at the foot of Caroline street, early in the morning of September 25 ('94) by Mr. E. Speddin, of the tugboat Chicago, and they were numerous until November 14 ('94, Patapsco Marsh). On March 8 ('94), between two and three hundred were bedded off Grace's Quarter, and on June 9 ('94), a bunch of six were still off Ocean City.

Fairly common on fresh water, this species has been taken as follows: On October 29, '92, two at Lake Roland (Fisher), and three from a bunch of five at Powhatan Dam (Gray). Several have been shot at Loch Raven (Dukehart), and on November 5, '93, a pair were on the Potomac near Knoxville (Fisher).

Chen hyperborea nivialis (169a). Greater Snow Goose.

"Along the Atlantic coast it may be considered rare" (Birds N. W., 549). On April 26, '80, Mr. Mitchell, of Cecil Co., presented a specimen to the Maryland Academy of Sciences; at Legoe's Point one was noted on March 10, '90, and early in October, '90, two flew over the bridge at Sparrows' Point, about 40 feet up (Fisher).

Anser albifrons gambeli (171a). American White-fronted Goose.

Rare straggler. One·shot on the Potomac in 1856 was bought in the Washington market for the Smithsonian (Smith. Rept., '56, 68).

On Nov. 12, '92, a young male was shot at Grace's Quarter, Baltimore County, by Mr. Charles D. Fisher, and by him presented to the Maryland Academy of Sciences. It was flying alone and came in to decoy's answering the usual goose call.

Branta canadensis (172).   Canada Goose.

Common winter resident. On October 4 ('94), four were seen sitting on a log at Spry's Island, by George B. Fowler, and on the 20th ('94), they were "exceedingly numerous in Day's Hollow on Gunpowder River; when they flew up it sounded like thunder; a few days later they had all left" (Edw. A. Robinson). They remain with us during winter in rather more limited numbers, and are again numerous in early spring, the latest record being April 10 ('90, Resler).

Inland, they are liable to be found anywhere in the state during spring and fall flights. On November 5 and 6 ('93, Fisher), quite a number were on the Potomac between Knoxville and Brunswick, and on April 15, 16 and 17, ('83, Small), the reservoir at Hagerstown "fairly swarmed with them."

In New Jersey "some linger as late as May 12" (Birds E. Pa. and N. J., 60).

Branta canadensis hutchinsii (172a).   Hutchin's Goose.

"South in winter through United States, chiefly west of the Alleghanies" (Manual, 117). "A goose, from its small dimensions and 16 tail feathers apparently referable to this sub-species, was taken at Cobb's Island, in the winter of 1888–89 by Capt. Crumb" (Birds Vas., p. 49).

Branta bernicla (173).   Brant.

Arriving and departing at about the same time as the common goose, this species is not so abundant. Most numerous during spring and fall migrations, quite a number as a rule winter with us, but I have no characteristic dates.

Branta nigricans (174).   Black Brant.

"Very rarely straggling to the Atlantic coast" (Manual, 118), this species has been taken on the New Jersey coast (Birds

E. Pa. and N. J., 60) and " Capt. Crumb informs me that it is usual for one or two to be obtained at Cobb's Island nearly every winter" (Birds Vas., 49).

Olor columbianus (180). Whistling Swan.

Common winter resident on the broad waters of tidewater Maryland, and during spring and fall flights liable to be seen anywhere in the state.   On September 26 ('93), one was shot on the Potomac near Weverton, by John Leopold.   On November 4 ('93, Fisher), several bunches were at the same place, while on April 15, 16 and 17, ('83), two were on the reservoir at Hagerstown (Small).

While swans are more or less difficult to shoot, they often " bed " on broad water out of range in large numbers.   On January 20, '94, I counted 82 standing on ice at the mouth of Gunpowder River, and a week later 194 on the water at the same place, where I am told they at times appear in greater numbers.

Olor buccinator (181). Trumpeter Swan.

Casual on the Atlantic coast.   "In Turnbull's list (Birds E. Pa.) this species is included on the authority of reliable sportsmen who have shot it on the Chesapeake Bay " (Birds N. W., 545).

In the Oologist, Vol. vi, 15–16, is quite an interesting article on the taking of one from a bunch of 12 or 15 on Slaughter Beach Marsh, Del., by Mr. G. L. Stevens, of Lincoln, Del., on November 9 ('89), at which time they were flying southward.

Order HERODIONES.—HERONS, STORKS, IBISES, ETC.

Family IBIDIDÆ—Ibises.

Guara alba (184). White Ibis.

Regularly north to the Carolinas in summer and casually to Long Island; two have been recorded from New Jersey and one from Pennsylvania (Birds E. Pa. and N. J., 61).

Plegadis autumnalis (186). Glossy Ibis.

" Warmer parts of the Eastern Hemisphere, also more southern portions of the Eastern United States" (Manual, 124). "At very irregular periods in the spring, small flocks have been seen on the coast of the Middle States, and on the Eastern Shore of Maryland and Virginia," and one taken near Baltimore, and two in the District of Columbia in 1817 are mentioned (Water Birds of N. A., i, 95–6).

Family CICONIIDÆ—Storks.

Tantalus loculator (188). Wood Ibis.

Regularly "north to the Carolinas, casually to Pennsylvania and New Jersey" (Key, 653). "The late Judge Libhart, in his ornithological report of Lancaster Co., Pa., says : "I obtained a fine specimen of this species shot from a troop of 10, by Mr. M. Ely, on the Susquehanna, in July 1862," and "Dr. A. C. Treichler, of Elizabethtown, has specimens in his collection which were shot in Lancaster Co. in the early part of July, 1883, shortly after severe storms" (Birds Pa., 53). "On July 2 1893, Mr. Fred. Zoller brought me 2 females, adult and young ; they were killed on the flats a short distance from the Washington monument, and on the Maryland side of the Potomac" (E. M. Hasbrouck, Auk, x, 92). Two were taken near Bloomery, Hampshire Co., W. Va., by Dr. A. Wall (American Field, xxii, 82).

Mr. Robert Shriver, of Cumberland, writes me : "About 30 years ago I shot a Wood Ibis. This was the only specimen I ever saw ; it was first seen by myself near the Potomac River bank, but before it was captured a dozen hunters were after it and it evaded them for several days. I always valued this specimen highly and am sorry it has gone 'the way of all flesh.'"

On October 15, 1893, Mr. Jacob F. Saylor described a bird to me taken during "wheat harvest" at the bend of the Gunpowder Falls in Dulaney's Valley, Baltimore Co.; from his description, and later corroboration by Mr. Dukehart, I am satisfied it can be no other than this species.

Family ARDEIDÆ—Bitterns and Herons.

Botaurus lentiginosus (190). American Bittern.

Fairly common during spring and fall, a few spend the summer with us and possibly breed. It may also stay over winter during mild seasons. From March 25 ('93, Gray) they are numerously noted until May 5 ('93, Resler), and in fall from September 1 ('91, Tylor), when one was taken at Tuckahoe Creek until October 10 ('94), when one was caught alive by Mr. Jacob Kirkwood early in the morning in front of No. 103 Elliott street in Baltimore City. This he kindly kept in a box until I examined it. It was an ordinary sized dark plumaged male.

Mr. J. E. Tylor supplies me with the following items: "Between the 20th and 30th of August, 1891, I killed a male Bittern in the Adkin's woods, one mile south of Easton, and mounted same. On the first day of September, 1891, Dr. E. R. Trippe, of Easton, in company with A. G. Pascault, of the same town, shot a male Bittern in Tuckahoe Creek, five miles below Hillsboro; this I also mounted. On July 14, 1894, on Hog Creek, Gunpowder River, I flushed one from the marsh, but did not secure it."

"Mr. William H. Buller, residing at Marietta, Lancaster County, Pa., in a letter dated July 29, 1889, addressed to me, writes as follows: 'I am inclined to believe that the American Bittern breeds in the vicinity of Schock's Mills, a few miles west of Marietta; while I have never found its nest, or seen its young, yet I have so frequently seen the bird in that vicinity during the summer, that I think it probable that it breeds there'" (Birds Pa., 55).

Dr. Coues, speaking of the District of Columbia, says: "Resident and rather common" (A. C., 100), and "I have procured it in January at Washington" (Birds N. W. 529). "Rather common from August to April at Washington" (Richmond).

Ardella exilis (191).　Least Bittern.

Common during summer, this little bird is seldom seen except by those who look for it, and owing to the marshes in which it lives, not always to be found by those who do.

On May 12 ('94, Wholey), six were noted in Patapsco Marsh, but that they are here earlier is shown by nests containing 5, 3 and 1 eggs, respectively, which were noted at the same place on May 17 ('93, A. Wolle), and in the last week of September, 1894, one was brought to A. Wolle, who kept it alive for some time in his shop window. The latest date for eggs is July 8 ('92), when four nearly fresh were collected at Sparrow's Point. It also possibly occurs more or less regularly on inland swamps. In May, 1893, one was caught alive in Dulaney's Valley by Mr. Dukehart.

Ardea herodias (194).　Great Blue Heron.

Between March 30 ('93, Gray) and November 19 ('90, Resler) this species is noted with more or less regularity, but so far I have been unable to find a "heronry" in Maryland.

In winter it has been observed at Back River on December 7 and 21 ('92, Resler). Noted at Hagerstown (Small), and Cumberland (Shriver).

Ardea egretta (196).　American Egret.

Irregular during late summer and early fall, this species has been noted from July 5 ('82), when one was at Gunpowder Falls a short distance below the Belair Road, to September 1 ('93 A. Wolle), when one was at Gunpowder Marsh; on September 23 ('94, Tylor) two were at Ocean City.

In Dulaney's Valley, one out of a flock of 9 was shot by Mr. Thos. Peerse in front of his house, and another was taken by Mr. Dukehart in the fall of 1893.

Possibly some may nest in Maryland, as "Mr. Wm. Palmer has known this species to nest in Arlington Cemetery, Va." (C. W. Richmond, Auk, v, 19), and "up to 1877 they bred near Townsend's Inlet, N. J. (Scott, B. N. O. C., '79), and a few may still breed in the state" (Birds E. Pa. and N. J., 63).

Ardea candidissima (197).   Snowy Heron.

Not as numerous as its larger relative, it probably visits us about the same time of year. On August 3 ('89, Resler), one of 6 was shot at Middle River, and on August 7 ('80, Resler), another at Back River. On August 10 ('88, Tylor), one at Choptank River, and later in same month another near Greensboro. On August 25 ('93), one near Marlboro, by Richard Cantler. At Washington it is given as "not uncommon about the marshes of the Potomac towards the end of summer and early fall" (A. C., 98). In 1886 they were breeding at Seven Mile Beach, N. J. (H. G. Parker, O. and O., iii, 138).

Ardea tricolor ruficollis (199).   Louisiana Heron.

" Warmer portions of Eastern North America ; north, casually to New Jersey" (Manual, 131). "Turnbull states that it has occasionally been taken on the New Jersey coast, but we can find no records of recent captures " (Birds E. Pa. and N. J., 64). " I have in my possession a skin of an immature bird taken several years ago at Cobb's Island, and understand from Captain Crumb that he has heard of 2 others that have been seen or taken " (Birds Vas., 50).

" A printed record (I cannot give reference) says one was preserved in the Maryland Academy of Sciences that was shot in Maryland" (Richmond). Mr. Uhler has very distinct recollections of the reception of this specimen, but cannot recall particulars ; it occurred in 1868 or '69.

Ardea cœrulea (200).   Little Blue Heron.

Rare straggler from the south; I know of but one specimen taken near Baltimore, this was shot at Day's Marsh, on October 7, 1892 (Fisher). Others no doubt have been taken and possibly some in the white plumage have been noted under Snowy

Heron.   At Washington, it is given as*" rare and only casual towards the end of summer " (A. C., 99), and "although usually rare, it is sometimes extremely abundant, a flock of about 150 frequenting the shores of the Potomac during August 1875" (P. L. Jouy, Field and Forest, iii, 51).

Ardea virescens (201).   Green Heron.

Common summer resident, arriving the last of March, a nest ready for eggs was found April 15 ('91), and as late as October 12 ('89, Resler), one was taken at Back River.   I have found eggs in the nest from April 22 ('81), to June 12 ('94); sets are 4 of 3, 13 of 4, and 8 of 5.   On several occasions, in different parts of the state, I have come across " heronries " of this species, the number of nests ranging from 6 to 17, but single nests may be found scattered everywhere, usually near marshy land, or water.

Nycticorax nycticorax nævius (202).   Black-crowned Night Heron.

Locally common, otherwise rare, from April 1 ('93, Gray) to October 17 ('94, Resler).   Only a few miles from Baltimore city a colony has nested for several years of which Mr. G. H. Gray, supplied the following :   " More than 6 years ago information of the approximate breeding site was received.   It was not until April 16, 1892, however, that it was found, when about 30 nests were in various stages of completeness.   They were in slender black oaks, near the top, and about 50 feet from the ground.   On April 30 the majority contained eggs.   2 had 5; 3 had 4, 10 or 12 had 3 ; the others 2 or 1.   On June 24, many of the young were perched about the limbs of the nest trees while others were still on the nests.   On March 25, 1893, none had arrived, but on April 1, seven were seen.   On May 6, nests with eggs were found in a clump of small scrub pines adjoining the oaks, which they had vacated.   A few days previous to our visit a severe wind storm had shaken these pines and the ground was strewn with broken egg shells. One nest however, had 5 eggs and they ranged down to 1 or more; while 2 nests were not yet completed. I measured the height of 11 nests and found they ranged from 36 to 49 feet from the ground. On June 9 they were nearly all back in the oaks occupying the old nests where they had birds just hatched, the few in the pines containing birds 10 or 12 days old. Among the oaks the shells of three eggs were found under 30 nests, the other 6 had 1 or 2, but possibly a few eggs were not yet hatched. The night of July 30 was spent in this heronry, but as the moon set early nothing could be seen. Each nest however, seemed to be visited by the parents about once every hour and the noise the young made was something wonderful.

"Owing to changes they did not nest here in 1894, but they were found in 1895 located about a mile off, as the crow flies, again in black oaks, the nests being from 42 to 48 feet up. On May 5 a few nests showed eggs, 4 and 5 being noted, while on May 11, 3 were noted with 4 fresh, one with 4 nearly fresh ; one with 5 nearly hatched, one with 4 young and a rotten egg, one with 3 and another with one fresh egg."

"Wm. Palmer has known this species to nest in Arlington Cemetery" (C. W. Richmond, Auk, v, 20). "Occasional in winter " (Richmond).

Nycticorax vialaceus (203).   Yellow-crowned Night Heron.

A few have been taken in Pennsylvania and New Jersey, where it is regarded as a "very rare straggler from the south" (Birds E. Pa. and N. J., 65).

"Rare summer visitant in the coast region; I have examined a young bird that was taken at Cobb's Island, and think that another bird has been taken there" (Birds Vas., 51).

Order PALUDICOLÆ.—CRANES, RAILS, ETC.

Family GRUIDÆ—Cranes.

Grus mexicana (206).   Sandhill Crane.

"Rare or irregular in the east " (Key, 667). "A specimen of this bird has been procured in the District of Columbia. We doubt, however, that the bird has been seen here alive for the past quarter of a century, and it might properly be retired from the active list" (A. C., 100).

Family RALLIDÆ—Rails, Gallinules and Coots.

Rallus elegans (208).   King Rail.

Fairly common summer resident of our fresh and brackish marshes. At Patapsco Marsh they have been taken from May 17 ('93) to October 6 ('76, Resler), while on October 13 ('94) I saw a small box of mixed game from Cumberland opened, among others it contained one King Rail.

At Tolchester, on May 30 ('91, Fisher), a nest containing 6 fresh eggs was found, and on June 15 ('91, Fisher), another with 10 fresh eggs at the some place.

Possibly some may winter during open seasons; vide—"Two King Sora were brought to Fredericksburg, Va., on Saturday (Jan. 28, '93), by Mr. Geo. Newton, of Stafford, and presented to Capt. M. B. Rowe. The appearance of these birds at this season is said to be quite remarkable, as they generally leave on the first appearance of frost" (Va. item in the *Sun* paper). Stafford is about 6 miles from the Potomac. In January, 1895, Mr. Scoggins received two King Rail "from the Rappahannock."

Rallus longirostris crepitans (211).   Clapper Rail.

Possibly resident in southern Maryland except when driven out by severe frost; this species, where not persecuted by pot hunters is fairly numerous on salt water marshes during summer. On May 17 ('93), 3 were shot on Patapsco Marsh, one, a female, contained a large number of eggs in the ovary, some quite large, and one in the oviduct already spotted and ready for extrusion. Noted at Hagerstown in October, '79 (Small).

Rallus virginianus (212).   Virginia Rail.

Common during migration, a few remain during summer. On July 8 ('92, Wholey) a specimen was taken at Sandy Point, near

Ocean City, and at Washington it has been "seen during the breeding season and undoubtedly breeds" (C. W. Richmond, Auk, v, 20).

Mr. Stone's remarks apply equally well to Maryland, he says: "Summer resident in fresh marshes, bogs and swamps along the coast, though apparently not in the true salt marshes except in migrations not very abundant inland, but rather plentiful along the New Jersey coast; arrives May 1st and remains until October 25, or occasionally later" (Birds E. Pa. and N. J., 66).

On March 20, '95, in market I saw one with a bunch of Wilson's Snipe; they were perfectly fresh and may have been shot near Baltimore, but I could not ascertain locality.

Porzana carolina (214).   Sora.

While this species migrates north regularly in spring, it is seldom noted, as the gunners are not then on the marshes, but during August, September and October, they are slaughtered by thousands on the marshes of tidewater Maryland.

On August 25 ('94, Fisher) they were abundant on Day's Marsh, they remain so until the first frost; this of course makes their going further south a variable date. Other dates are few, so they are given in full. On November 18 ('90, Fisher) one was shot in Somerset County. Single birds were taken on December 26 ('90) and January 22 ('95), at the mud hole back of "Sonny" Barranger's in Canton, by Mr. Jas. Holton, and on April 3 ('93) one was shot in a marsh a short distance north of Chincoteague Island, and consequently near the southern Maryland line, by Dr. R. H. P. Ellis, of Baltimore city.

On July 25, '93, four birds were flushed from a small piece of cattail, by Mr. Geo. Todd, close to his house on North Point. This seems to point to their possible breeding here in limited numbers. It is given as breeding in Chester and Lancaster Counties, Pa. (Birds Pa., 71), and at Washington it has been "seen during the breeding season and undoubtedly breeds" (C. W. Richmond, Auk, v, 20). Under date of May 25, '95, Mr.

Richmond says: "I doubt very much now whether it ever breeds here, but it is common in August. Birds have been shot here, one on November 8, '78, by Peter Burger, one on November 9, '78, by S. F. Baird, and one in March, '75."

Inland, several were shot during the fall of '93, in Dulaney's Valley (Dukehart); at Hagerstown, in October, '79, and September, '80 (Small); at "Cumberland during all of April (not seen after May 1, as shooting stopped then), and from August to October 15, '94" (Zacharia Laney).

Porzana noveboracensis (215). Yellow Rail.

"Eastern North America, not abundant, very secretive" (Key, 674), and as the bird is small it is no wonder it is not often seen; possibly it may yet be found to breed with us. On April 27, '93, one was shot on Patapsco Marsh by Richard Cantler, this I saw. On May 18, '89, at Hog Creek Marsh, Harford County, one was flushed twice but not secured, by Mr. W. H. Fisher, and on October 20, '94, one was presented to me in the flesh; it was received with a mixed lot of birds, in a box sent from Back River Neck by a market gunner.

"In the collection of the National Museum are two Yellow Rails, both of which were taken on the marshes of the Potomac River near Washington, the first by T. E. Clark, October 4, '79, the second by A. S. Skinner, March 28, '84" (H. M. Smith and Wm. Palmer, Auk, v, 147).

Porzana jamaicensis (216). Black Rail.

"Not often found in the United States, being one of our rarest birds" (Key, p. 674). One secured at Piscataway, Maryland, was presented to the Smithsonian Institute by John Dowell, of Washington, D. C., (Smith. Report, '84, 145). "One seen in the District of Columbia during September, '61, but not secured. One taken 2 or 3 years ago is now in the Smithsonian" (A. C., 101). Several are recorded from Pennsylvania and New Jersey, (Birds E. Pa. and N. J., 67).

Ionornis martinica (218).   Purple Gallinule.

"South Atlantic and Gulf States, resident, north casually to New England" (Key, 676).   Capt. Crumb has one mounted that came ashore on Cobb's Island during a storm in May '91, and was captured in the light house yard (Letter to W. H. Fisher). It has also been taken in York County, Pa. (Birds Pa., p. 73).

"One was seen in Centre market (Washington, D. C.) on August 24, '89 by Geo. Marshall. I visited the market, (Golden's stand) to see about it. The man in charge remembered the 'purple bird,' but thought it had been sold, he said it came from 'down the Potomac somewhere'" (Richmond).

Gallinula galeata (219).   Florida Gallinule.

Possibly a regular, though rare, migrant. At Stemmer's Run, Baltimore Co., one was taken on May 8, ('89, Resler). At Wasington, "on April 19, '92, Mr. Fred Zeller brought me a Florida Gallinule; while the species has been taken here before, this is the first specimen existing in collections. A few days later, about the 22d, Mr. J. D. Figgins secured a specimen from Frederick, Md., and on August 12, '92, Mr. Zeller brought me a young male" (E. M. Hasbrouck, Auk, x, 92).

" Regular transient, but not very common, most of the specimens secured in Pennsylvania and New Jersey have been taken on the Delaware and Susquehanna Rivers. Occurs May 1–10, and September 8 to October 20 " (Birds E. Pa. and N. J., 68).

Fulica americana (221).   American Coot.

Common migrant, noted from March 14 ('93, Fisher), at Gunpowder Marsh, to May 7 ('93), when a bunch of 5 were on the broad water of Chester River, and again from September 20 ('79, Resler), at Back River, to November 3 ('91, Resler), at Patapsco Marsh.

A number have been shot at Loch Raven (Dukehart). At Hagerstown, on April 16, '83, there was a remarkable flight of

these birds, hundreds being killed; they were also noted during
September and October, '79, and October, '81 (Small).

Order LIMICOLÆ—Shore Birds.

Family PHALAROPODIDÆ—Phalaropes.

Crymophilus fulicarius (222).   Red Phalarope.

"A species of circumpolar distribution during summer"
(Key, 614). "South in winter to Middle States" (Manual,
144). "A young bird taken on the Eastern Branch of the
Potomac near Washington, D. C., by Mr. F. S. Webster, on
October 17, '85, is now in the National Museum" (Smith.
Report, '87, 603).

Phalaropus lobatus (223).   Northern Phalarope.

Circumpolar, like the last species, but coming much further
south in winter. Under date of May 28, '95, Mr. C. W. Richmond writes me from Washington, "one was taken here in
September, '91, by Mr. Thomas Marron off Navy Yard bridge;
the specimen is now in the National Museum."

Family RECURVIROSTRIDÆ—Stilts.

Himantopus mexicanus (226).   Black-necked Stilt.

Uncommon on the Atlantic coast from Florida to Maine; now
rare, it may yet occur on the ocean front of Maryland. "Formerly it bred regularly in Cape May County, N. J., and on
Egg Island, Delaware Bay (Turnbull, '69). We know of no
recent captures of this species" (Birds E. Pa. and N. J., 70).

Family SCOLOPACIDÆ—Snipes, Sandpipers, etc.

Scolopax rusticola (227).   European Woodcock.

Straggling from Europe, this species has been taken quite
close to Maryland, but as far as I know not within the state.
"In the early part of November, '86, Mr. D. N. McFarland,

of West Chester, Pa., killed a large female in the 'barrens' of
East Nottingham Township, Chester County, Pa." (Birds Pa.,
78), and Dr. Coues reports one being shot in Loudoun County,
Va., in '73 (Forest and Stream, vi, 180).

Philohela minor (228).   American Woodcock.

Resident, except when frozen out, and fairly common in spite
of the June and July gunners who slaughter many birds before
they are much more than half grown, and leave others still
younger parentless.

Nesting dates range from March 30 ('80), when four slightly
incubated eggs were found by Mr. W. L. Amoss near Falls-
town, Harford County, to July 4 ('93), when a nest, also with
four eggs, was found (all sets I know of are four).

At Cumberland, where they are not found in winter, the first
was shot on March 28, some years ago, and the last on
December 12, '94 (Zacharia Laney).

Gallinago delicata (230).   Wilson's Snipe.

Common during migrations.   On February 24, '95, I flushed
one from a warm spring in Dulaney's Valley, and on March 9
('95, Henninghouse) they were numerous at Gunpowder Marsh,
where they were still numerous on April 22 ('92, Pleasants),
the last spring date being April 29 ('94), in Dulaney's Valley,
though at Washington they are noted until May 5 (Richmond).

In the fall I have them noted from September 20 ('93), at
Patapsco Marsh, to December 11 ('94), Bush River.   There is
every possibility that more or less remain with us during mild
winters.

The following note on this species breeding in Maryland is
taken from Lewis' *American Sportsman*, (1885 edition, p.
244).   " In the month of May, 1846, while wandering in company with Mr. E. Lewis over his extensive estate in Maryland,
we sprang a Wilson's Snipe from the midst of an oat field, and
being surprised as well as attracted by its singular manœuvres,

we made search for its nest which we soon found with four eggs in it. The situation selected for incubation could not have been better chosen in any portion of the country, as it was on a rising piece of ground, with a southern exposure, protected in the rear by a large wood, and at the foot of the high ground was a considerable extent of low marsh meadow watered by a never-failing stream, along the border of which the anxious parent at any time could obtain a bountiful supply of food."

Mr. Zacharia Laney, of Cumberland, informs me that he has taken this species from February 28 to the last of the gunning season, April 30 ; how much later they stay he does not know, but some years ago while exercising a pair of young dogs in June, they flushed a pair from a marshy slew.

Macrorhamphus griseus (231).   Dowitcher.

Common during migration in tidewater Maryland during April and May, and from early in August to the first touch of cold weather; a specimen in the collection of Mr. A Resler was taken at Back River as early as March 6 ('75). Inland, Mr. Dukehart has secured a number in both spring and fall in Dulaney's Valley.

Macrorhamphus scolopaceus (232).   Long-billed Dowitcher.

Of this western species "seven were killed from a flock on the Anacostia River, D. C., in April, 1884, by a gunner who sold them in the market for Jack Snipe. One similar to the others was secured and mounted by one of the writers and has been identified by Mr. Ridgway as the western species" (Hugh M. Smith and Wm. Palmer, Auk, v, 147).

Micropalama himantopus (233).   Stilt Sandpiper.

Casual migrant on the Atlantic coast, one was " taken on the Patuxent River, Md.,September 8, '85, by Mr. H. W. Henshaw. This capture was made beyond the regular District of Columbia boundary, but was, however, included in what has been tacitly regarded as its faunal and floral limit" (Hugh M. Smith, Auk, iii, 139).

Tringa canutus (234).   Knot.

Common migrant, given on the New Jersey coast from May 15 to June 1, and from August 15 to September 15 (Birds E. Pa. and N. J., 73). On Cobb's Island 19 were shot on May 20 ('91, Fisher), and from May 14 to 28 ('94), they were " quite numerous occurring in large flocks. On May 25 hundreds of these birds were seen feeding along the extensive mud flats on the outer sea beach" (E. J. Brown, Auk, xi, 259).

On August 19, '93, quite a number were shot a few miles south of Ocean City on the beach (Janon Fisher).

Tringa maculata (239).   Pectoral Sandpiper.

" United States chiefly during migrations, when observed in wet grassy meadows, muddy ponds, flats, etc." (Key, 626). "Rare in spring, not uncommon from September 25 to November" (A. C., 96.)

Near Washington, " on April 22, '88, two were shot by W. F. Roberts; several were taken on August 3 (about '89); at St. George's Island, Md., several were seen on September 3 and 11, '94; and one was shot October 22, '60 " (Richmond).

" Regular migrant on the New Jersey coast in April, and from the middle of August to the first of October; occasional in the interior " (Birds E. Pa. and N. J., 73).

Tringa fuscicollis (240).   White-rumped Sandpiper.

" Transient on the New Jersey coast, but not very common, associating with the Least Sandpiper, and arriving and departing with it " (Birds E. Pa. and N. J., 73).

"E. J. Brown has two or three skins taken in May, between the 15th and 24th, '94, at Smith's Island, Va." (Richmond).

Tringa bairdii (241). Baird's Sandpiper.

Migrant, "rare on the Atlantic coast, common in the interior" (Key, 626). "One was shot by R. S. Matthews, near Four Mile Run, Va., on September 3, '94, and Wm. Palmer shot a second specimen on September 25, '94, at the same place" (Richmond).

Tringa minutilla (242). Least Sandpiper.

Common migrant all through May, and again from August 2 ('92), [in New Jersey, Stone says July 15] to November 3 ('94), when four were shot at Back River. Numerous along the shores of tidewater Maryland, bunches of "Peeps" may be found along all our rivers, streams, runs, ice ponds, etc., wherever there is wet mud.

Tringa alpina pacifica (243a). Red-backed Sandpiper.

Common during migrations in tidewater Maryland; this species has also been noted on our larger inland waters; possibly some remain during mild winters in southern Maryland. On September 3 ('93), one was at Loch Raven, and on the 17th, two. On March 13 ('92, Wholey) four were at Waverley, while on May 24 ('93), I shot two out of a bunch of four on Hail Point, Kent County, at the mouth of Chester River.

Ereunetes pusillus (246). Semipalmated Sandpiper.

Common during migrations, but not as numerous as *T. minutilla*, with flocks of which it is generally found, arriving and departing at the same time.

Near Washington, D. C., recorded from August 13 ('94) to October 26 ('87), and again in May (Richmond).

Ereunetes occidentalis (247). Western Sandpiper.

"During the last week of August, 1885, I found the western bird quite as common as the eastern at Piney Point, St. Mary's County, Md., on the Potomac River. If anything, the former was the most numerous, for, of the 18 specimens of

Ereunetes preserved, 14 were identified by Mr. Ridgway as *occidentales*, and these too were taken at random from a lot of about 25 dead birds" (Hugh M. Smith, Auk, xi, 385). Several were shot at Virginia Beach, on September 6 and 7, 1884, by Messrs. Henry Seebohm and C. W. Beckham (Auk, xi, 101). Two have been taken in New Jersey: one on September 14, '80, the other on May 17, '92 (Birds E. Pa. and N. J., 75).

Calidris arenaria (248).   Sanderling.

Abundant coastwise during migrations; several were seen and one shot at Cobb's Island on May 20 ('91, Fisher), and one was taken at Ocean City on September 23 ('94, Tylor).

"Near Washington one was taken on September 23, '94, by Wm. Palmer; another in '74, and one on October 24, '85, at Gravelly Run" (Richmond).

"Abundant transient on the New Jersey coast, keeping pretty much to the beach, April 18 to June 1, September 14 to October 15. Some also are said to remain through the winter. Dr. W. L. Abbott secured one specimen in the spring migration as late as June 13. Occasional on the lower Delaware" (Birds E. Pa. and N. J., 75).

Limosa fedoa (249).   Marbled Godwit.

This species "does not appear to go far along the Atlantic coast northward" (Key, 635). Rather rare transient on the New Jersey coast, where it seems to have been more plentiful formerly; occurs in May, and again from the last week of July to September 15. Dr. Warren states that a few have been captured in recent years in Lancaster, Philadelphia and Delaware Counties" (Birds E. Pa. and N. J., 75).

Limosa hæmastica (251).   Hudsonian Godwit.

"Much less abundant in the United States than the preceding, and appears to range chiefly along the Atlantic coast" (Birds N. W., 494). "On May 16, '86, I shot a Hudsonian Godwit at

West River, Md., in a grass field adjoining the village of Hales-ville" (J. Murray Ellzey, Forrest and Stream, xxvii., 264).

Totanus melanoleucus (254).   Greater Yellow-legs.

Common migrant, but not so numerous as the following spe-cies.   On March 26 ('75, Resler) one was taken at Patapsco Marsh, and on June 7 ('94) one at Ocean City.   In fall, Mr. Stone gives them as early as July 15 in New Jersey (Birds E. Pa. and N. J.), and Mr. Richmond at Washington from July 25, but my earliest note is a flock of 35 in Dulaney's Valley on August 12 ('94), while as late as November 4 ('93), I received one from Patapsco Neck (probably shot about the 2nd or 3rd), and another from Cumberland, also probably shot at the same time.

Totanus flavipes (255).   Yellow-legs.

More numerous than the former, in the spring from March 15 ('95), at Havre de Grace, to May 17 ('93, Wholey), at Patapsco Marsh, and again from August 12 ('94), at Loch Ra-ven, to September 5 ('93, J. H. Fisher, Jr.), at Spring Gardens. At Washington, September 11 ('94, Wm. Palmer and R. S. Matthews).

Totanus solitarius (256).   Solitary Sandpiper.

A regular but not abundant migrant, generally found singly or in pairs in spring, and in small flocks in fall.   Noted from April 25 to May 30 ('91, Gray), and from August 13 ('93, Gray) to October 12 ('89, Resler).

This species is a rare summer resident, though its nest has not yet been recorded in Maryland.   On July 14, '93, one was in company with a Killdeer in the bed of Gwynn's Falls, at Cal-verton (Gray and Blogg).

"Occasionally one is seen during the breeding season" at Sandy Springs (Stabler).   " In Maryland and Virginia.   .   .   . I have shot birds in August so young as to leave no doubt in my mind that they were bred in the vicinity" (Birds N. W., 499).

At Washington, D. C., it is recorded as follows: "April 26 ('91) one seen, May 16 ('88) one seen, July 20 ('90) one noted, July 28 ('89), two seen, August 8 (94, Wm. Palmer) one seen, August 11 ('89) one seen, and on August 21 ('94 E. J. Brown), several seen and shot" (Richmond).

"Dr. Treichler, of Lancaster County (Pa.), mentions it as an irregular breeder; he has found young about half grown in the Conowingo meadows early in July" (Birds Pa., 91).

On May 23, '93, I came across one feeding in an ice pond with but little water ; I was within 20 feet and watched it through a field glass. It waded with a dainty step, sometimes having perceptibly to pull its feet out of the mud, and once when it got in deeper water swam a few feet with a hurried stroke. Small insects on or in the water, on bottom, on weed stems or on bank it swallowed at once. Tiny tadpoles it worked a second between mandibles and dipped a couple of times in water, apparently to get them head first. One large one it immediately ran ashore with and hammered it on the ground for some little time before swallowing it. In wading it sometimes had the water up to its breast and belly. Generally not more than the bill was immersed, but often the head, occasionally the neck, and once half of the body.

Symphemia semipalmata (258).   Willet.

While not as numerous as it used to be, this species is not uncommon on our ocean front, where it still breeds in limited numbers. At Chincoteague Bay, I noted one on June 5 ('94); another on the next day, and three on the 7th at the same place ; and Mr. C. W. Dirickson, of Berlin, says: "On their way north in spring a few stop and stay with us all summer. They lay their eggs in very much the same place as the terns, and in fact you can sometimes find both nesting very close together." I am informed that they breed in large numbers on Chincoteague Island, and also on Mockhorn and Smith's Islands, Va. "A few still breed on the New Jersey coast" (Birds E. Pa. and N. J., 76–77).

On August 19 ('93, Janon Fisher) a number were shot a few miles down the beach from Ocean City.   On August 27, '93, a bunch of about 30 were on the flats opposite the Navy Yard, Washington, D. C. (B. A. Bean, Forest and Stream, xli, 230).   On November 3, '94, I saw a single bird in the Baltimore market, still quite fresh, which had been shot "down the necks," possibly one or two days before. '

Pavoncella pugnax (260).   Ruff.

On September 3, '94, a bird of this species was shot at Four Mile Run by Wm. Palmer (Richmond).   This European species has occasionally been taken on the coast of New England and the Middle States (for references see Key, 641).

Bartramia longicauda (261).   Bartramian Sandpiper.

Common during migrations and fairly represented in summer. In Dulaney's Valley they were first noted on April 21 ('94, Fisher), the migratory birds leaving about the middle of May; they are noted again from August 11 ('95) to September 8 ('95).   This year ('95) two pair spent the summer in the north end of the valley, and about a mile apart.   These I visited weekly; up to July 14 both birds of each pair would allow of quite close approach, one, presumably the female, often circling round within 50 feet.   They gave every sign of nesting, but it was not until early in August that the young were seen flying with the parents.   On August 8 two young birds were shot. These birds were flying in two bunches of 4 and 6 until September 1.

On July 3 ('95) one, evidently lost, was flying round calling over an electric light in Baltimore City at 11.30 P. M.

"On July 27 ('89) one was shot at Laurel by Geo. Marshall. First recorded at Washington on April 6 ('92); few stop on their passage, but they are commonly heard while migrating at night" (Richmond).   "A summer resident, rare at that season" (A. C., 83).

Actitis macularia (263).   Spotted Sandpiper.

Common summer resident from April 8 ('93, Gray) to Oct. 18 ('94). Numerous all through the state wherever there is water; they regularly spend the summer at Druid Hill Lake, and a few years ago I frequently observed a pair on Jones Falls opposite Union Station, where no doubt they had a nest.

On May 30 ('91, J. H. Fisher, Jr.), a nest was found at Tolchester, containing four nearly hatched eggs, and on July 21 ('95), four young were still being led by the parents.

On July 21 ('95), I watched one for some time; the speed with which it ran after a fly, with sudden doubles, was remarkable; every once in a while it stopped to scratch the back of its head, finally it thought a wash would do that head good, so flying to a shallow part of the run, it squatted down in the water and began to duck its head under, scratching the back of its head on both sides with the nail of the long middle toe, until it got every feather raised and quite wet. The balance of the body received no attention. Hopping on a stone it dried its head by rubbing it against its sides several times and then flew away. This head washing lasted nine minutes.

Numenius longirostris (264).   Long-billed Curlew.

Breeding on the Atlantic coast as far north as North Carolina, and casually north to New England, this species is "a straggler on the New Jersey coast, occurring generally in May and September" (Birds E. Pa. and N. J., 78). On May 23 ('93), two were on Hail Point, Kent County. On August 19 ('93, Janon Fisher), quite a number were shot a few miles south of Ocean City, and during September ('93, G. A. Rasch), they were unusually plentiful at Cobb's Island.

Numenius hudsonicus (265).   Hudsonian Curlew.

Migratory through the United States. One day in August, 1881, W. H. Fisher shot three about eight miles below Ocean City, Md. On May 19 ('91), he saw a good many at New

Marsh, Cobb's Island, but they would not decoy, and two days later he saw quite a number at Smith's Island. Given as "common transient along the New Jersey coast, occurring May 1 to June 1, and July 15 to September 15" (Birds E. Pa. and N. J., 78). On May 10 ('95), five and one Curlew flew over Baltimore City, presumably they were of this species.

Numenius borealis (266). Eskimo Curlew.

Migrating through United States. Captain Crumb calls this species a rare and irregular migrant at Cobb's Island (Birds Vas., 57). "Rare transient on the coast, appearing in May and again in September, according to Turnbull" (Birds E. Pa. and N. J., 78).

Family CHARADRIIDÆ—Plovers.

Charadrius squatarola (270). Black-billed Plover.

"Migratory in United States, preferably coastwise, common, but less so than *dominicus*" (Key, 598). Three were noted at Waverly on May 12 ('94, Wholey), and one in Dulaney's Valley the next day. On May 19 ('91, Fisher) three were shot at Cobb's Island, Va., and next day two more from numerous flocks observed. Given as common on the New Jersey coast from April 30 to May 22, and from latter part of July to September 15. Dr. W. L. Abbott has taken specimens on June 3 and November 7 (Birds E. Pa. and N. J., 78–79).

Charadrius dominicus (272). American Golden Plover.

In New Jersey "very erratic transient, rarely seen in spring, but at irregular intervals occurring in large flocks in the fall" (Birds E. Pa. and N. J., 79). At Washington, "rare and irregular migrant" (Richmond); at Cumberland (Shriver).

Ægialitis vocifera (273).   Killdeer.

Common in summer all over the state; this species is very numerous during migrations in tidewater Maryland, where during mild winters more or less may winter.   In Dulaney's Valley this species is a common bird from March 10 ('95) to December 2 ('94), while at Powhatan Dam it was noted as early as February 25 ('93, Gray).   The set of eggs is, as far as I know, four.   Fresh eggs were noted May 28 ('92), one hatched and three pipped, on June 5 ('91), and young still with the parents on July 23 ('93).

On and after July 14 ('95), they are usually to be found in flocks, 53 on one occasion being counted in a close bunch.

. Ægialitis semipalmata (274).   Semipalmated Plover.

Common migrant appearing on ocean front, sand beaches of Chesapeake Bay and mud flats of our larger inland waters from early in May to the 29 ('80, Resler), when one was taken at Patapsco Marsh.   Returning late in July, and remaining until September 22 ('94, Tylor), when five were taken at Ocean City.   On August 12 ('94) and 29 ('93, Fisher), they were numerous along Loch Raven.

Ægialitis meloda (277).   Piping Plover.

On June 5 ('94), one feeding in the wash of the waves a few miles from Ocean City was noted and three days later, one back where the sand and marsh grass meet, gave unmistakable evidence of having either eggs or small young, but neither could be found.

Given from April 15 to May 15, and in September and October as a transient, a few breeding, and also wintering on the New Jersey coast (Birds E. Pa. and N. J., 79–80).   Capt. Crumb states that it has nested at Cobb's Island.

Ægialitis meloda circumcincta (277a). Belted Piping Plover.

This western species is occasional on the Atlantic coast. On May 3, '84, a specimen, now in the National Museum, was obtained on the shore of the Potomac opposite Washington, near the Long Bridge (H. M. Smith and Wm. Palmer, Auk, v, 147).

Ægialitis wilsonia (280). Wilson's Plover.

Common during summer along the Atlantic coast as far north as Virginia. At Cobb's Island, Va., Mr. H. B. Bailey found it a comparatively common bird, May 25–29, '75 (Auk, i, 26), and they were found breeding there on May 20, '91 (Fisher). "Rare straggler on the New Jersey coast, where it probably bred a few years ago" (Birds E. Pa. and N. J., 80).

Family APHRIZIDÆ—Turnstones.

Arenaria interpres (283). Turnstone.

Common migrant on our ocean front during May, August and September. Apparently irregular on the Chesapeake, they have been taken as far up the Potomac as Washington, D. C. "Three Turnstones in the National Museum were taken in the District of Columbia by Mr. C. Drexler in 1860 (?). In June, '82, Mr. J. A. Moore killed a bird at Jones Point, Va., near Washington. In May, '81, Mr. O. N. Bryan secured one, and saw another at Marshall Hall, Md., and we know of the occurrence of three others on the Potomac River within the past three years" (H. M. Smith and Wm. Palmer, Auk, v, 147–148).

Family HÆMATOPODIDÆ—Oystercatchers.

Hæmatopus palliatus (286). American Oystercatcher.

Common on the coast of the Southern States during summer. At Cobb's Island they were found breeding, and eggs were taken during June, '88 (Theo. W. Richards, Oologist, vii, 186), and on May 20, '91, several pairs were observed there (Fisher). On June 5, '91, one was noted flying south at North Beach, a few miles south of Ocean City, Md. "Very rare straggler on New Jersey coast" (Birds E. Pa. and N. J., 81).

Order GALLINÆ—GALLINACEOUS BIRDS.

Family TETRAONIDÆ—Grouse, Bob-whites, etc.

Colinus virginianus (289). Bob-white.

Common, resident; May 2 ('93) is the earliest date I have heard the well-known call of "Bob-white," this is generally stopped by the early part of September; it has, however, been heard as late as October 23 ('88, Wholey). Their equally well-known whistle and the faint "click-click" of the covey may be heard at any time. Usually the call is given from the top of the fence, but I have heard it repeatedly given by birds in trees. The nest is seldom found except during harvest when numbers are uncovered. Sets are one each of 8, 12, 14, 18 and 22, and 2 of 10. A nest with eggs was found as late as September 1 ('95). A covey of 7 or 8 holding together rather late was flushed near Magnolia on May 4 ('93, Fisher).

Coturnix communis—European Quail.

In the fall of '79, Messrs. Poultney, Trimble & Co. imported about 200 birds from Italy. These Mr. Charles D. Fisher turned loose on his place at Ruxton, but never saw anything of them afterwards. The following spring about 1000 were imported, and Mr. Kleibacker tells me there were several eggs in the boxes when they arrived. Quite a number were turned out on the "Dundee Shore," where occasionally one or two were seen during the summer, but early in fall they entirely disappeared. Messrs. Charles B. Rogers and Geo. Brown also liberated a number in Green Spring Valley, where at least one pair nested. Mr. Rogers writing me under date of January 27, '93, says: "There were several pairs of the birds on our property. I remember that a nest was found near our blacksmith shop, it was on the ground in a cluster of weeds, and if I remember correctly had 13 eggs in it. Nothing however was ever seen or heard of the Quail after the first winter." Mr. Isaac Slingluff writing in reference to the above, says : "The nest was found close to the bank of Shoemaker's Run. I think there were 12 or 13 eggs in it, 4 of which did not hatch." One of these is now in the possession of Mr. Jesse Slingluff.

Mr. E. W. Hasbrouck, of Washington, writes me, "In February, '95, I saw a bunch of European Quail exposed for sale on the street ; they were said to have been killed near Opequan Creek, but I could get no further information."

Bonasa umbellus (300).   Ruffed Grouse.

Common, resident. On May 2 ('94, Fisher), a nest with 10 fresh eggs was found. During July a covey of young, about as large as partridge, were seen near Ellicott City by Mr. Basil Sollers, and on June 10, '95, at Vale Summit, I came across a pair with 8 or 10 young about the same size.

Tympanuchus americanus (305).   Prairie Hen.

Early in the winter of '85 or '86, Col. Edw. Wilkins got 12 or 15 birds from the west and kept them until spring when he turned them loose in his orchard, on the Chester River, about four miles below Chestertown, Kent County ; a few days later two or three were seen and then they disappeared. On Eastern Neck Island, Mr. Spencer Wicks shot one in the fall, and Mr. Newton Bogle several times saw another dusting itself in the road in the front of his house. Possibly these were some of Col. Wilkins birds. "Mr. Ridgway records the killing of a Prairie Hen on the Virginia side of the Potomac, near Washington, March 17, '85 (Forest and Stream, xxiv, 204 and 248). It has been suggested that it was a descendant of birds, liberated previously at Snow Hill, Maryland" (Birds Vas., 59).

Family PHASIANIDÆ—Pheasants, etc.

Phasianus colchicus.   European Pheasant.

Col. W. F. Mason McCarty tells me that some years ago a number were liberated on the grounds of the Woodmount Gunning Club in Washington County. They are now fairly numerous, and this colony may be considered as established. Occasionally birds wander off, and have been shot quite a distance away from the preserve.

Meleagris gallopavo (310).   Wild Turkey.

Mr. Robert Shriver writing from Cumberland, says, "Indigenous here, about as abundant as ever, they seem to be less numerous some seasons, owing probably to severity of weather or excessive hunting."

Mr. Wm. H. Fisher supplies the following: "Mr. J. R. Ridgley tells me that on his farm in Howard County, about 10 miles from Ellicott City, some years ago he was listening for squirrels and became conscious of a wild turkey, when too late to secure it." He also says "that now ('93) there are a few near his farm in a tract called Beaver Dam Woods."

Mr. Harden, of Georgetown, shot two in the winter of '81-82, between Georgetown and Tennallytown (A. C., '92).

"While at Weverton on September 26, '93, John Leopold told me that about the first of the month a flock of at least 10 were seen, and that now a flock of seven young, with two adults, is on the other (Virginia) side of the river. Occasionally he has seen turkeys fly across the river, and once one gave out and fell into the water " (Fisher).

Order COLUMBÆ—PIGEONS AND DOVES.

Family COLUMBIDÆ—Pigeons and Doves.

Ectopistes migratorius (315).   Passenger Pigeon.

Originally occurring in large numbers, but only occasionally seen of late years. On August 27, '93, I flushed a pair from a fence in the upper end of Dulaney's Valley and further down a single bird from a tree top. These birds I watched for some time through a field glass, but none of their actions differed from those of doves. Mr. Wm. H. Fisher supplies the following: "For about 10 days, in October '78, flocks of Wild Pigeons flew over our house at Mount Washington between 7 and 7.30 A. M.; 6 to 10 flocks of from 5 to 20 birds each day. In September, '88, I shot one near Bradshaw, and in September, '89, another in Dulaney's Valley; this last was flying with a flock of Doves. Mr. J. R. Ridgley tells me that he saw a flock of 50 or 60 about 8 miles from Ellicott City on September 17, '93. Clarence Cottman says he saw a flock of about 40 pass near the head of Lake Roland and fly up Green Spring Valley on September 19, '93."

Dr. Coues, speaking of the District of Columbia, says: "I once killed a specimen so newly from the nest as to cause me to believe that it had been hatched in the vicinity" (Birds N. W., 389).

They were once very common at Cumberland, but of late years have become very rare (Shriver). At Vale Summit I was told that the last flight occurred there on the evening of New Years day '77, when the sky was black with them and large numbers were killed.

Zenaidura macroura (316). Mourning Dove.

Common, resident. The usual set of 2 eggs is recorded from April 1 ('82, Small), at Hagerstown, to August 17 ('93, Stabler) at Sandy Springs; near Baltimore, from April 9 ('93) to August 13 ('90, Resler). The nest is usually placed on a fork, or among twigs on a horizontal branch, where it is flat and shallow, but I found one built in the fork of a split cedar 14 inches from bottom to top. I have also found a number in old nests (Robin, Purple Grackle, Cardinal, etc.), also on fence rails and one on top of a stump. Dr. Warren cites several on the ground (Birds Pa., 114).

During fall they unite into bunches and flocks, these are recorded from August 3 ('95) to April 3 ('93); as a rule they do not range above 25, but I have seen about 50, and Mr. Wm. H. Fisher saw one of at least 200 on August 26, '93.

Columbigallina passerina terrestris (320). Ground Dove.

Common in the Southern States; "its usual range is limited by the Carolinas, but I have a record of the capture of a specimen many years ago at Washington" (A. C., 91. Birds N. W., 390). Another specimen shot by Mr. Thos. Marron on Oct. 14, '88, at Broad Creek, Md., is now in the National Museum (Smith. Report, '89, 117, 358 and 801).

One was taken in Lancaster County, Pa., in '44 (Birds E. Pa. and N. J., 80).

Order RAPTORES—Birds of Prey.

Family CATHARTIDÆ—American Vultures.

Cathartes aura (325).   Turkey Vulture.

Resident, common., 14 sets of 2 eggs are recorded from April 19 ('82, W. L. Amoss, Fallston) to May 30 ('91, Stabler, Sandy Springs). I have several times heard of 3 eggs, but never could verify the statement. The eggs, nest there is none, are placed in hollow prostrate logs, hollow stumps, under rocks, stones, or bushes, and in one case under the worn side of an old straw stack. Given as resident at Hagerstown (Small), but only as casual at Cumberland (Shriver); at Vale Summit I only saw 3 in 10 days (June 5 to 14, '95).

While usually not noticed by other birds, I saw a crow chase one on May 8, '92, and a Fish Hawk chase another on May 30, '93.

Catharista atrata (326).   Black Vulture.

"North, regularly to North Carolina, irregularly or casually to Maine, New York, etc." (Manual, 222). "Rarely breeds North of 36°" (Bendire, 165). "On March 30, '95, at Kensington, Md., Mr. J. D. Figgins saw 4 birds which he supposed were of this species. He had never seen the Black Vulture in life, but was familiar with the other large birds known to occur here and from his description of the actions of the birds, I have no doubt they were really of this species" (Richmond).

Family FALCONIDÆ—Falcons, Hawks, Eagles, etc.

Elanoides forficatus (327).   Swallow-tailed Kite.

"On the Atlantic coast its natural limits appear to be the lower portions of Virginia, but it has more than once occurred in the Middle States" (Birds N. W., 332). On April 5, '93, I examined a mounted specimen; on inquiry I was told it had been shot by Mr. W. T. Levering, Jr., on Maidens Choice Lane, close to Kenwood Station, near Catonsville, Baltimore County, late in the summer (late July or early August) of '89.

While in Queen Anne County (May '92) my boatman several times mentioned a "white Hawk with a split tail" which he had observed occasionally, but he could not tell how often or at what time of year. Mr. A. P. Bowen writes me that it is occasionally seen in Prince George's County.

Circus hudsonius (331).   Marsh Hawk.

Common resident in tide-water Maryland, this species is common in the uplands from August 4 ('95) to May 2 ('91, Gray), and on June 29 ('92, Gray) one was seen near Powhatan. Mr. L. D. Willis, of Church Creek, Dorchester County, informs me that on a salt marsh of the Blackwater River, about 10 miles south of Cambridge, he found a nest of the Marsh Hawk containing 6 fresh eggs on June 2, '95. Next day the female was shot and an egg ready for extension was found in the ova-duct. The nest, a slight affair of dry reeds and grass, was placed on a slight elevation about 60 yards back from the open water and entirely surrounded by thick reeds.

Accipiter velox (332).   Sharp-shinned Hawk.

Common resident, but not very numerous in summer. On May 20 ('91, Blogg) a set of 4 eggs was taken, and on May 29 ('92, Fisher) another of 5. At Sandy Springs on May 16, '91, two fresh eggs, and 15 days later a set of 5 (Stabler).

For two years in succession this terror of the poultry yard nested close to our house, but as far as I could learn the chickens were not molested. In '84 the young had left the nest and were being fed round the house on June 8, 15 and 22. In '85 I spent May 31 and June 7 watching the young being fed. They were in a natural cavity of a chestnut tree in full view of the house and not 100 feet away from it. The entrance was a small rotted-out branch hole about 4 inches in diameter and about 40 feet from the ground. By means of a field glass I saw that the young were being fed on grasshoppers. On the approach of a parent they made a great noise, and as this could easily he heard all over the house, the feeding was continuous from "the dawn's early light," until it was too dark to see the birds come and go. The nest of this species is usually placed in the twigs of a cedar or pine.

Accipiter cooperi (333). Cooper's Hawk.

Resident, but at all times less numerous than *velox*. The nest is built early in April, and the birds stay around it a long time before the eggs are laid. On April 29 ('91, Stabler) a set of eggs was taken at Sandy Springs, and on June 20 ('88, Resler), two birds, at most a week old, were in a nest in Howard County. Sets are 1 of 2, 2 of 3 and 2 of 5.

Accipiter atricapillus (334). American Goshawk.

Maryland appears to be the southern limit of this species in winter, but it is not taken here often. In '68, Dr. J. Lee McComas, of Cumberland, sent two, shot in Maryland, to the Smithsonian (Smith. Report, '68, 57), and another sent to Dr. A. K. Fisher, was taken in a steel trap by Mr. Leizear, of Sandy Springs, on December 27 ('87, Stabler).

Buteo borealis (337). Red-tailed Hawk.

Resident, common, but not often seen in summer. Two slightly incubated eggs were taken on March 25 ('94, Hoen), and three birds just hatched were seen on May 7 ('93), while two birds apparently just out of the nest were trying to follow their parents on July 5 ('94). Sets are 1 of 1; 5 of 2; 2 of 3 and 1 of 4. Outside the breeding season they are more or less gregarious, at times being seen in quite large flocks.

Buteo lineatus (399). Red-shouldered Hawk.

Inside the limits of Baltimore City this is the most numerous hawk at any time of the year, but in Baltimore County *borealis* far outnumbers it. On March 20 ('93) two sets of fresh eggs were taken (a ring of snow was round each nest, and it was two feet deep on the ground); while on May 6 ('94, Wholey) a set nearly incubated was collected, and on May 31 ('91) a second set was taken. Sets are 5 of 2; 4 of 3, and 4 of 4.

Outside the breeding season, this species, like *borealis* may sometimes be seen in flocks, more often in bunches of 3 or 4 to 10, but generally singly or in pairs.

Buteo latissimus (343). Broad-winged Hawk.

Resident, but not common. On April 27 ('91) a set of three eggs was taken; on May 19 ('92, Blogg) a set of two, and on May 23 ('92, J. H. Fisher, Jr.) a set of three nearly incubated. At Sandy Springs, a set of three was taken April 9 ('91); one of three in May ('92); one of two on May 15 ('92), and another of two on May 22 ('92, Stabler).

The three *Buteos* are the hawks usually shot by our farmers, because they are large, fly slow, and are called "hen-hawks," while the much smaller, swift flying, *Falcos* and *Accipiters*, that may at times take chickens, escape.

Archibuteo lagopus sancti-johannis (347a). American Rough-legged Hawk.

"Along the Delaware, below Philadelphia, it is still found in considerable numbers from November to the end of March" (Birds E. Pa. and N. J., 87). It does not appear to be numerous in Maryland.' On January 24, '92, in Dulaney's Valley, one sitting on a tree allowed me to drive slowly past within twenty feet of it. Dr. M. G. Ellzey says this species was very numerous in Howard County during the winter and spring of '87–88 (Forest and Stream, xxxii, 212). At Sandy Springs one was shot by Mr. Leizear on March 17, '88 (Stabler). One was seen on the Virginia side of the Potomac, opposite Washington, on December 29, '79, by Mr. H. W. Henshaw (A. C., 88–9), and it has been taken in the District of Columbia (L. M. McCormick, Auk, i, 397).

Aquila chrysaëtos (349).   Golden Eagle.

An irregular winter visitant.   On June 30, '83, a mounted specimen was presented to the Maryland Academy of Sciences, by Dr. Murdoch ; it was shot at Back River, and had five toes instead of four.   On November 28, '94, one was shot at Otter Creek, Harford County; this specimen I had mounted.   Four were secured in Maryland, near Washington (see Smith. Reports, '62, 58; '69, 55; '75, 73 ; '91, 793).   On December 8, '87, one was shot at Gaithersburg (Fisher's Hawks and Owls, 97). One in the First National Bank of Cumberland was shot by Mr. Robert Shriver about 30 years ago.   "Until about 1856, for many years a pair is said to have nested in the southern part of Lancaster County on a lofty jutting cliff over the Susquehanna River" (M. W. Raub, Auk, ix, 200).

Haliæetus leucocephalus (352).   Bald Eagle.

Resident, and generally dispersed along the shores of the Chesapeake Bay and other larger waters of Maryland, being fairly common.   At Loch Raven it is no unusual sight to see one or more, and Dr. Wilson, of Glenarm, tells me that about 40 years ago there was a nest used year after year within half a mile of his house.   I have only found the nest "down the necks" and on the Eastern Shore; but they breed all along the Potomac as far up as Hancock.   On March 8 ('94) three eggs nearly fresh were taken; on March 27 ('95), two nearly fresh ; on March 29 ('93), two about two-thirds incubated ; on April 15 ('93), two birds, five or six weeks old, and on June 2 ('92), two birds about ready to leave the nest.   On February 26 ('93), two eggs were collected at Mount Vernon by E. M. Hasbrouck.   The nest being used for years, sometimes attains considerable size, eight feet across the top and seven feet high, is the largest one I have measured.

In confinement at Toledo, Ohio, two eggs were laid, incubation commenced on March 26, '86, and one bird was hatched on April 26 ; on March 18, '88, she again started to incubate two eggs; one bird was hatched on April 22, and another the next day (Henry Hulce, Forest and Stream, xxvi, 327 and xxx, 289).

Falco peregrinus anatum (356).   Duck Hawk.

"Universally, but irregularly distributed in North America, scarcely to be considered common anywhere, breeds as far south as Virginia at least, usually in mountainous regions" (Key, 536). "Nests sparingly from 35° north" (Bendire, 292). On March 5 ('87, Fisher), one was seen at Grace's Quarter, but it kept out of range. On December 10, '94, one was brought alive to Baltimore by a countryman, and lived for about four months in the window of No. 208 E. Baltimore street.

"Mr. W. T. Roberts got a female on November 16, '79, at Potomac Landing, near Washington" (Richmond). "Mr. Jouy states that the Duck Hawk has been known to breed at Harper's Ferry" (A. C., 87). In reference to this, Mr. W. H. Fisher had a conversation with a resident of Harper's Ferry on October 10, '93. He was not acquainted with the "Duck Hawk," but stated that the "Rock Hawk" nested on the face of the Maryland Heights, and that the site of the nest could be easily located by the "white-wash" after the young were hatched.

This species is reported as regularly breeding along the Susquehanna in Lancaster and York Counties, Pa. In the latter county a set of four slightly incubated eggs was taken on April 7, '80, by Mr. Geo. Miller (Birds Pa., 137).

Falco columbarius (357).   Pigeon Hawk.

"Whole of North America, breeding chiefly north of the United States" (Manual, 250). "Doubtless breeds in the mountainous portions of some of the Southern States" (Bendire, 299).

I have but few notes on this powerful little hawk. On April 19, '92, one was shot at Washington, that had been feeding on a Sparrow-hawk (Fisher's Hawks and Owls, 113). On April 26, '93, one was shot near Bay View by A. Wolle, it had been feeding on a Snow-bird. On April 30, '93, in Dulaney's Valley two birds swished close past me making for a small scrubby thorn bush, round which they went several times; finally one darted into its centre, the other kept round once more and lit on top. I then saw that they were a Catbird and a Pigeon Hawk, the hawk seeing me flew, but the Catbird remained apparently thoroughly exhausted. On October 30, '92, I surprised one feeding on a dove. A not uncommon migrant at Washington (Richmond). "Mr. F. L. Washburn, of Johns Hopkins University, has reported to the Agricultural Department that he observed several pairs, apparently breeding at Harper's Ferry, April 12, '87" (Birds Vas., 62).

Falco sparverius (360).   American Sparrow Hawk.

Resident, common, but most numerous during the migrations of small bird. Hence, on March 18 ('93, Wholey and Gray) about 75 were seen in Dulaney's Valley. During summer they are not often seen except in the locality of the nest. Dates for eggs range from April 17 ('95, Henninghouse) to May 28 ('93), and a second set was taken June 24 ('94). On August 4 ('95) flying young were still being fed by the parents. Sets are 9 of 4 and 6 of 5.

Pandion haliaëtus carolinensis (364).   American Osprey.

Common summer resident on all our larger waters, extremely numerous on the arms of the Chesapeake. On March 11 ('93, Blogg) five were seen at Fulton Avenue. On the Eastern Shore it is maintained that they always arrive on St. Patrick's Day, March 17. Dr. Sharp's attention was called to this, and he reported that the first bird arrived at Rock Hall on March 14, '95. Late in September most have gone south, but they have been noted later, the last on November 8 ('92, Blogg).

On April 3 ('93) they were patching up old nests at the mouth of Gunpowder, and on April 24 ('94, Tylor) sets of fresh eggs were collected in Talbot County, where on August

19 ('83) I noted a number of nests with young still in them. Sets are 4 of 1, 9 of 2, 19 of 3, and 3 of 4.

Inland, this bird is seen more or less regularly at Loch Raven, and one was seen as far up the Gunpowder as Cockeysville on April 17, '94 (Fisher). Several were over the Potomac, at Brunswick, on September 26, '93 (Fisher).

Family STRIGIDÆ—Barn Owls.

Strix pratincola (365).   American Barn Owl.

"Not abundant north of the Carolinas" (Birds N. W., 300). "At Washington, where the Barn Owl is by no means rare, they begin nesting from the last week in April to about May 10" (Bendire, 327). "The National Museum collection contains two eggs of this bird taken from the Smithsonian towers, one in June, '61, the other June 1, '65" (C. W. Richmond, Auk, v, 20). On June 28, '90, seven half-grown young were found in this tower (Fisher's Hawks and Owls, 136). "On December 8, '93, a young bird that had but recently left its nest was caught, probably hatched some time in October. On February 27, '95, another of about the same age was picked up in a bush in the Smithsonian grounds. This was certainly not over two months old, and must have been hatched in the latter part of December, if not early in January ; certainly a most unusual time of the year for this owl to breed in this latitude" (C. E. Bendire, Auk, xii, 180–81).

Occasionally one is secured anywhere in tidewater Maryland, and they seemed to be more numerous than usual during the spring of '93. On April 6 one was shot at the Old Marine Hospital ; on the 20th, a male at Aberdeen, Harford County; on the 22nd, another male on Patapsco Neck, and on July 27th a female and five downy young were taken alive near the Old Marine Hospital, by A. Wolle.

Family BUBONIDÆ—Horned Owls, Hoot Owls, etc.

Asio wilsonianus (366).   American Long-eared Owl.

Nocturnal in its habits; this species is resident, but as far as
I know not common, though a locality may yet be found where
it is numerous. On April 22 ('93, Gray and Blogg), a set of
six eggs, nearly hatched, was collected near Randalstown, from
an old crow's nest about 20 feet up a small pine.
"Common resident at Washington" (Richmond); "at Hagerstown one was shot in January" ('79 Small).

Asio accipitrinus (367).   Short-eared Owl.

"It is more than likely that it breeds, occasionally at least,
in suitable localities along the borders of the extensive marshes
of the sea coast of the southern Atlantic States; by far the
greater number, however, breed north of the United States"
(Bendire, 332). "Decidedly the commonest owl about Washington, especially in winter" (Birds N. W., 307), where it has
been taken in November, January, March and April, and at
Sandy Springs it has been taken in December, January, February and March (Fisher's Hawks and Owls, 148–9). On April
1 ('92, Fisher) one was taken at Ruxton. On October 22 ('92,
Gray), one at Pikesville; on November 8 ('90, Pleasants), one
at Towson, and on December 2 ('93, R. C. Watters), one in Dorchester County.

Syrnium nebulosum (368).   Barred Owl.

Resident, numerous "down the necks," otherwise fairly common round Baltimore. March 16 ('95, Fisher), one fresh egg,
and May 5 ('94, Fisher), two young birds, four or five days old,
are extreme nesting dates. Sets are 2 of 1, 4 of 2, and 1 of 3.

Nyctala acadica (372).   Saw-whet Owl.

While by no means common in winter, a number have been
recorded. On November 13 ('92, Blogg) one flew into a
brightly lighted room in the city. On December 23 ('75,

Resler) one was taken alive, and on March 4 ('94, Tyler and Fisher) another was taken alive at Bird River. The records of the Maryland Academy of Sciences show one presented on May 15 ('79), possibly a mounted specimen.

Quite a number are recorded from the District of Columbia and adjoining portions of Maryland. F. W. Webster notes eight obtained early in October in different years (Auk, iv, 161). One taken on November 1 ('78); one on February 12 ('59), and one on March 12 ('89) (C. W. Richmond, Auk, vi, 189). One on December 12 ('90); three on January 4 ('91), and one taken alive in the Smithsonian on February 4 or 5 ('91) (E. M. Hasbrouck, Auk, viii, 313); one on November 1 ('89) (Fisher's Hawks and Owls, 162), and another at Ivy City, on December 3 ('89) (W. A. Merritt, Oologist, viii, 313).

Megascops asio (373).   Screech Owl.

Common resident and generally dispersed, but more often heard than seen. I have heard them all through winter at Waverly, Baltimore City, where on January 16, '92, at 7.30 P. M. one was "laughing" as merrily as in June, though there was 6 inches of snow on the ground. Nesting dates range from April 4 ('92), 4 fresh eggs, to June 4 ('93), 3 birds just hatched, while on July 24 ('93) young nearly grown were still being fed by the parents. Sets are 2 of 2, 5 of 3, 9 of 4, and 3 of 5. As far as I know the gray and red phases of plumage are about equal round Baltimore.

Bubo virginianus (375).   Great Horned Owl.

Common resident, but most numerous in heavily wooded sections of the state, especially in tidewater Maryland. On February 11 (94, Tylor) two fresh eggs were taken on the Virginia side of the Potomac, near Alexandria. On February 25 ('95, Tylor), two eggs, one-third inculated, at Magnolia. On April 2 ('93), one bird about 2 weeks old at Bush River, and on April 12 ('93, A. Wolle), 2 eggs "down the necks."

In Dulaney's Valley on March 10 ('95) a nest was found with an egg about to hatch and a young bird not 24 hours old, and on April 24 ('92) two young birds just out of the nest were captured alive.

Sets are 1 of 1, 8 of 2, 1 of 3, and early in April '91 four birds just out of the nest and sitting together on a limb were seen in Talbot County, two of these were captured and raised in confinement (Tylor).

Nyctea nyctea (376). Snowy Owl.

More or less numerous in Maryland during severe winters, this species cannot at any time be called common. Quite a number are on record from all over the state, but I cannot get exact dates.

Order PSITTACI—PARROTS, ETC.

Family PSITTACIDÆ—Parrots and Paroquets.

Conurus carolinensis (382). Carolina Paroquet.

Originally well known in tidewater Maryland, the only occurrence for many years is recorded as follows: "In September, '65, while gunning for Sora on the Potomac River, Mr. Ed. Derrick fired into a flock of strange birds flying overhead, killing several which proved to be Carolina Paroquets. He had one mounted and kept the specimen in his house for a number of years. Other parties on the marsh at the same time shot numbers of the birds" (H. M. Smith, and Wm. Palmer, Auk, v, 148).

Order COCCYGES—CUCKOOS, KINGFISHERS, ETC.

Family CUCULIDÆ—Cuckoos, etc.

Coccyzus americanus (387). Yellow-billed Cuckoo.

Common summer resident, but more often heard than seen. Extreme dates are April 28 ('88, Resler) and October 14 ('91, Resler). Nests with eggs have been found from June 7 ('93,

Fisher) to August 23 ('91). As far as I have been definitely able to note sets, they are 6 of 2, 3 of 3, and 2 of 4. I have found nests with all the eggs in the same state of incubation and others with various stages, from large young birds to eggs in different stages of incubation. At Vale Summit they were fairly common, on June 9 ('95) a nest contained 2 fresh eggs.

Coccyzus erythrophthalmus (388).   Black-billed Cuckoo.

A common migrant and rare summer resident. May 8 ('89, Resler) and May 21 ('93, Wholey) seem extremes of the spring movement, and August 4 ('95) to October 3 ('93, Wholey) cover the fall. On September 28 '92, (Wholey) "they were the most numerous birds seen in the woods."

On July 7, '93, (Gray) a nest with one young bird and 3 nearly incubated eggs was found at Calverton, while further out the Franklin Road, in an overgrown corner, two broods were raised the same year.

At Washington it is noted as "rare from May 2 to the middle of October" (Richmond). At Hagerstown, noted in July, August and September ('80), and from May 11 to September ('81, Small). At Vale Summit, on June 14, '95, I found the nest, containing 2 eggs about one-half incubated, of the only pair there.

Family ALCEDINIDÆ—Kingfishers.
Ceryle alcyon (390).   Belted Kingfisher.

Abundant during spring, summer and fall; a number winter in tidewater Maryland during mild seasons, only leaving when frozen out. The nest tunnel, in a bank, preferably but not always over water, is remarkably close to 4 inches in diameter and usually about 5 feet long, though I have seen two of not quite 2 feet and one of over 10. As a rule they go straight in but occasionally they make a bend. The nesting hole at the end is rounded in the shape of a flattened sphere and averages 16 inches across by 8 inches high. In this, the first egg is laid on the bare ground, but by the time the eggs hatch about a quart of loose fish scales and bones have accumulated. May 9 ('92, Blogg), 5 fresh eggs, and May 31 ('92), one fresh egg, are extreme dates; full sets are 3 of 6, and 6 of 7.

Order PICI—WOODPECKERS, ETC.

Family PICIDÆ—Woodpeckers.

Campephilus principalis (392). Ivory-billed Woodpecker.

However the distribution of this species may have been, it is now very restricted. Audubon says (iv, 124): "On the Atlantic coast North Carolina may be taken as the limit of its distribution, although now and then an individual of this species may be occasionally seen in Maryland."

Dryobates villosus (393). Hairy Woodpecker.

Resident, but not common, and, as it is usually found in heavy timber, appears much rarer than it really is. On June 2 ('94, Fisher) a very noisy pair evidently had a nest, but it was not found owing to lack of time. Young of the year were noted on July 10 ('92) and on August 20 ('93). On May 8 ('95, H. C. Oberholzer) a pair were feeding young near the Great Falls of the Potomac, on the Maryland side.

Dryobates pubescens (394). Downy Woodpecker.

Common resident. Nests with eggs have been noted from May 4 ('91) to May 22 ('93); the set being 5. On June 8 ('84) young were nearly ready to leave the nest, and on July 4 ('93) young not long out of the nest were seen.

Dryobates borealis (395). Red-cockaded Woodpecker.

"Pine swamps and barrens of South Atlantic and Gulf States, north to Pennsylvania" (Key, 481); "irregularly north to New Jersey" (Manual, 283). Dr. Ezra Meichner in his Catalogue of Chester County Birds, published in 1863, writes, "accidental, very rare" (Birds Pa., 167).

Sphyrapicus varius (402).   Yellow-bellied Sapsucker.

Common during migrations, September 27 ('79, Resler) to October 26 ('94), and again from March 12 ('92, Gray) to May 2 ('93, Fisher). No doubt this species winters in southern Maryland, as specimens have been taken near Baltimore on November 8 ('84, Resler); November 12 ('92, Gray); November 26 ('93); December 6 ('93, J. H. Fisher, Jr.); December 24 ('92, Blogg), and January 1 ('92, Resler).

At Washington, "J. D. Figgens got one on January 14, '88; one was shot February 15, about '59, and one was seen about the middle of January, '94" (Richmond).

At Hagerstown they were noted during January and December, '79; January, February, March, October, November and December, '80, and from January right along to July, and also in October, '81 (Small). On July 6, '95, (Tylor) adults feeding young were noted at Deer Park.

Ceophlœus pileatus (405).   Pileated Woodpecker.

Fairly common in the heavily wooded parts of Maryland. Early in June '95 a nest was found by Mr. L. D. Willis near Church Creek, Dorchester County. It contained 3 eggs; 2 nearly hatched, the other rotten. It measured 2 feet 2 inches deep by 8 inches in diameter, the entrance was 5½ inches across and 20 feet from the ground, in a rotten stub. November 17 to 22, '94 (Fisher), quite a number were seen in Somerset County, and one was observed to enter a hole in a stub.

"Mr. Palmer has 3 specimens bought in market (Washington) on the 9th of January, '79, which had been shot in Maryland, near the District line" (A. C., 81–2). "Said to be not rare at Johnson's Gully, Maryland, near Marshall Hall, and about 14 miles from Washington. Must be quite common in Virginia, as a market gunner brought me nine at one time. He would not disclose the locality, but said it was in Virginia" (Richmond). Quite a number are exposed for sale in our Baltimore markets each winter, but they are all said to come from Pennsylvania. On November 7, '91, I saw one that had been shot in Carroll County.

"The Indian hen used to be common round Cumberland, but is now very rare" (Z. Laney).

Melanerpes erythrocephalus (406). Red-headed Woodpecker.

This very erratic species, common one year in a certain locality and the next entirely absent or only in limited numbers, is resident, migratory or anything else, apparently at its own sweet will. During the severe winter of '92-3 Mr. W. N. Wholey and I had 8 or 10 birds that we went out regularly to see. They were resident in localities about one-half mile apart and stayed all winter; one was very noisy, the others were silent. The following winter none remained. May 3 ('91) and June 19 ('92) are extremes for eggs. Sets are 3 of 5, 1 of 4, and 1 of 3. As a rule they dig their own holes, telegraph poles being often used, but on May 3, '91, I found three fresh eggs in a hollow log leaning against a fence. The entrance was 15 inches from top to bottom and 4 inches across, while the cavity only went down 6 inches.

At Washington "not very common and local. It usually spends the winter in smaller numbers, or else keeps more secluded" (Richmond).

Melanerpes carolinus (409). Red-bellied Woodpecker.

This southern species is not common around Baltimore. Occasionally noted from August 3 ('87, Resler) to May 16 (91, Blogg). "Rather common near Laurel, where it is a permanent resident" (Richmond). At Washington it is given as "a permanent resident, rare" (A. C., 83); "very rare, Mr. Henshaw saw an individual about the last of May, '87" (C. W. Richmond, Auk, v, 21). In Queen Anne's County they were quite numerous on March 4 and 5, '93, though there was 12 inches of snow on the ground, and the thermometer went down to 8° during the intervening night.

At Princess Anne, Somerset County, they were rather common from November 13 to 22 ('94, Fisher), and Mr. E. G. Polk writing from there, says: "They stay here all summer, at least until after cherries are ripe, as I have shot a number out of the trees, where they were stealing cherries."

Colaptes auratus (412).   Flicker.

Common migrant, more numerous during migrations. April 28 ('94, Tylor), in Talbot County, and May 12 ('95), near Baltimore, to June 23 ('93), are extremes for eggs. Sets are 1 of 2, 3 of 5, 4 of 6, 2 of 7, 3 of 8, 1 of 9, 1 of 10, and 1 of 11. Flickers usually dig their own holes, but they will nest in natural cavities and various other places; if undisturbed, using the same site for years.

Order MACROCHIRES—Goatsuckers, Swifts, Humming-birds, etc.

Family Caprimulgidæ—Nighthawks, Whip-poor-wills, etc.

Antrostomus carolinensis (416).   Chuck-will's-widow.

North, regularly in summer to North Carolina; in a letter to Wm. H. Fisher, Capt. Crumb says he has taken this species at Cobb's Island, Va.

At Odenton, Anne Arundel County, upon two occasions in July, Prof. P. R. Uhler has observed single birds of this species, their size making them quite conspicuous among the Whip-poor-wills which were quite numerous there.

Antrostomus vociferus (417).   Whip-poor-will.

Locally common during summer, it was first heard April 9 ('93, Wholey), and on April 27 ('93, Fisher) they were common, remaining so until September 20 ('91), the last being noted September 28 ('94).   At Washington, "to October" (Richmond).

On May 5 ('94, Tylor) one fresh egg was found, and on June 10 ('91, Fisher) one young bird a few days old.

Chordeiles virginianus (420). Nighthawk.

About equally numerous and just as local as the Whip-poor-will, from April 29 ('93, J. H. Fisher, Jr.) to October 15 ('92, Blogg). At Washington, April 20 (Richmond). In fall, large flocks migrating are noted from August 29 ('94) to September 25 ('92, Wholey). Nesting on the bare ground, this species has discovered an excellent substitute in the flat roofs of the houses of Baltimore City, where a large number breed. On June 8, '91, one young bird a few days old was seen within stone's throw of the City Hall, and on July 31, '94, two young birds were flying after their parents.

Family MICROPODIDÆ—Swifts.

Chætura pelagica (423). Chimney Swift.

Common summer resident from April 16 ('91) to October 4 ('93); extreme dates are March 30 ('95, Blogg) and October 17 ('88, Resler); at Hagerstown, April 6 to October 16 ('80, Small); at Cumberland, April 5 ('95, Z. Laney).

On May 15 ('92) a pair were seen mating, but they were not noticed breaking twigs until June 3 ('94); eggs were in nests on June 19 ('81), and young birds fell down the chimney on August 27 ('93). Sets are 2 of 5.

Family TROCHILIDÆ—Hummingbirds.

Trochilus colubris (428). Ruby-throated Hummingbird.

Common summer resident; first noted at Washington on April 28 (Richmond) and numerous around Baltimore from May 1 ('92, Resler) to September 25 ('92); the latest date is October 3 ('88, Resler), when one was perched on a telegraph wire in the city, and ('90, Wholey) when one was taken. The usual set of two eggs is noted from June 2 ('93, J. H. Fisher, Jr.) to July 10 ('92, Blogg).

Order PASSERES—Perchingbirds.

Family Tyrannidæ—Flycatchers.

Milvulus forficatus (443).   Scissor-tailed Flycatcher.

"A *Milvulus* probably *M. forficatus* is given in the original edition as having been obtained by Mr. C. Drexler, on May 6, '81. We have never been satisfied of the accuracy of the information, even supposing veracity on the part of our informant, and in our remarks on Mr. Jouy's list, we spoke as if inclined to drop the species from the list; but we have no more authority for doing so than for retaining it, so we make no alteration in a record which, unfortunately, must always remain dubious" (A. C., 75–6). One "was sent to the Smithsonian Institute by Mr. R. B. Taylor, of Norfolk, Va., . . . shot on January 2, '82 in his door yard in that city" (Robert Ridgway, Auk, viii, 59). Under date of May 5, '93, Capt. Crumb states that he has secured a specimen at Cobb's Island.

Tyrannus tyrannus (444).   Kingbird.

Common summer resident from April 14 ('95), to September 15 (93, Gray). On September 23, '91, I saw a single bird 2 miles north of Martinsburg, W. Va., and at Hagerstown they are noted from April 13 ('83) to October ('79, Small). During migrations they appear in flocks, sometimes over 100 being together, these have been noted from April 30 ('93) to May 3 ('93, Fisher), and from August 4 ('95) to September 7 ('94). Extreme dates for eggs are May 31 ('93) and July 17 ('92). On August 12 ('94), young were still being fed. Sets are 1 of 1, 1 of 2, 13 of 3, and 3 of 4.

Tyrannus verticalis (447).   Arkansas Kingbird.

"This is a western species added to the list in '77, by Mr. Jouy, who found it in the flesh in market September 30, '74. In point of fact, it was not actually got in the District, but in some adjoining portion of Maryland. There is no doubt about the bird, as the specimen is preserved in the United States National Museum. (Smith. Report, '74–5, 32. Jouy's Catalogue, '77, 5 and 11)" (A. C., '76).

Myiarchus crinitus (452).   Crested Flycatcher.

Common summer resident from April 29 ('93, Gray), to September 2 ('93, Gray); extreme dates are April 27 ('93, Gray), and September 21 ('94). Eggs are recorded from June 6 ('91), to July 9 ('93). Sets are 1 of 3, 1 of 4, 4 of 5, and 2 of 6.

Sayornis phœbe (456).   Phœbe.

Common from March 18 ('94) to October 17 ('94), quite a number of single birds have been observed during the remainder of the year. This year ('95), however, they have been absent; a few were observed between March 31 and May 13, after which none were seen until September 15.

Extreme nesting dates are April 8 ('94), a nest ready for eggs and July 8 ('94), eggs nearly hatched. Sets are 3 of 3, 10 of 4, 13 of 5, and 1 of 6.

At Hagerstown, under date of January 26, '82, Small says: "Has probably been with us all winter, was seen December 8 and 26, and January 18, 19, 20 and 21."

Contopus borealis (459).   Olive-sided Flycatcher.

Rare migrant. "The claim of this species to a place in our list (of the District of Columbia) rests upon Mr. Ridgway's observation near Fall's Church, Va., where several birds were noticed in September, '81. Further west in Virginia, the species cannot be considered very rare, individuals having been observed for 3 or 4 successive summers by one of the authors, and Dr. A. K. Fisher has taken a specimen in the Bull Run mountains" (H. M. Smith and Wm. Palmer, Auk, v, 148).

"Judge Libhart states that this species (probably 15 or 20 years ago) was found as a breeder in Lancaster County, where, however, in recent years it has been observed by Dr. Treichler only as a rare spring and fall migrant" (Birds Pa., 194).

Contopus virens (461).   Wood Pewee.

Common summer resident; first noted on April 25 ('85, Resler), when one was taken, and numerous on May 8 ('94). A
number were noted October 15 ('93), and two days later ('83,
Resler) one was taken. Nesting dates range from June 8 ('92,
Blogg), a set of fresh eggs to September 12 ('92, Wholey)
young birds not able to fly far. Sets are 1 of 1, 5 of 2, and 7
of 3.

Empidonax flaviventris (463).   Yellow-bellied Flycatcher.

Migrant, not common. Spring notes are few. On May 14
('93), a male was taken and 4 days later another, 3 days after
which 5 were seen (Wholey). On May 16 ('83), and 19 ('93),
single birds were taken (Resler). During fall they are frequently recorded between August 31 ('93, Gray) and October 6
('94).

At Washington, a "spring and fall migrant arrives first week
in May and we have seen it in the fall until the third week in
September. One shot July 28," (A. C., 78) "migrant during
whole of May, and August and September" (Richmond).
"Tolerably common migrant near Washington, generally in the
scrub pines" (Dr. A. K. Fisher, Birds Vas., 67). This last
observation may account for its not being more frequently noted
near Baltimore, as but a small percentage of our local observations have been made among the pines.

Empidonax virescens (465).   Acadian Flycatcher.

First noted on April 30 ('93, Wholey), and common from
May 11 ('94,) to August 24 ('94 and '95,) the last was recorded
on September 11 ('94). On May 31 ('91,) a nest was ready for
eggs and on July 30 ('93, Fisher), one contained eggs ready to
hatch. Sets are 28 of 3, and 1 of 4.

Empidonax traillii alnorum (622a).   Traill's Flycatcher.

Rare migrant. On May 5, '93 (Resler), one was taken at Back
River, and on May 11, '93, (Wholey)another at Waverly. "One

was taken by the writer on May 13, '88, at the Potomac River, Alexander County, Va.; another by Mr. Ridgway on May 18, '88, at Laurel, Maryland, and a third by myself on May 19, '88, in Virginia, opposite Georgetown. Several others were subsequently seen and observed" (Wm. Palmer, Auk, vi, 71), "This has always been regarded as the rarest of the flycatcher's; very few having been taken up to the present year ('92). On and about May 18 for several days they were quite common and a number were taken" (E. M. Hasbrouck, Auk, x, '93). "Common at times, I saw several on May 23, '91. Wm. Palmer shot one on May 10, '94, and another on August 27, '89, and I shot one on September 17, '90" (Richmond).

Empidonax minimus (467). Least Flycatcher.

Rare migrant, specimens have been taken on April 29 ('93, Gray), on April 30 and May 7 ('92, Pleasants), on September 11 and 25 ('93, Gray), and September 28 ('92, Resler). At Washington, "common from April 25 to May 25 and from August 28 to September 25 " (Richmond).

On Dan's mountain, June 5 to 14, '95, three or four pairs were mating in a grove of heavy timber.

Family ALAUDIDÆ—Larks.

Otocoris alpestris (474). Horned Lark.

Irregularly abundant in flocks, which are sometimes quite large, from November 10 ('94, Gray), to March 19 ('92, Gray); they are most numerous in tidewater Maryland. At Washington, from the first of November to April (A. C., 41).

Otocoris alpestris praticola (474b). Prairie Horned Lark.

This sub-species appearing in company with *alpestris* can only be identified by close comparison with a series of specimens. One, of several shot at Powhatan, on February 25 ('93, Gray), was pronounced of this sub-species by Mr. Ridgway.

"Two in the collection of Wm. Palmer have been identified by Mr. Henshaw as belonging to this race; they were taken in February, '81, and were in company with numbers of *Otocoris alpestris*. On February 16, '88, eighteen specimens of this variety were taken by Wm. Palmer, near Washington, from a flock of 50 or 60 birds that had been noted in the vicinity throughout the winter. About half a dozen other specimens have recently been obtained by various collectors" (H. M. Smith and Wm. Palmer, Auk, v, 148).

Family CORVIDÆ—Crows, Jays, etc.

Cyanocitta cristata (477).   Blue Jay.

Common resident. A pair were mating on April 13 ('93), and eggs were collected, 5 on May 13 ('82) and 4 on May 30 ('94, Tylor).

Corvus corax principalis (486).   Raven.

Now rare in Maryland, but, without doubt, may be credited to our ocean front and also to the mountains of Western Maryland. During Christmas week '92, about 20 were seen at Bayard, W. Va., but they could not be approached within rifle range. On December 6, '93, several were seen at the same place (J. H. Fisher, Jr.). Bayard is within 5 miles of the Potomac. During July, '80, Ravens were found on Cobbs, Boone and Mockhorn Islands (Robert Ridgway, Auk, vi, 118). Recorded from Franklin, Somerset and York Counties, Pa., (Birds Pa., 202). One was taken at Hagerstown in October, '80 (Small).

Corvus americanus (488).   American Crow.

An abundant resident; scattered over the country in summer, and gathering into large "roosts" in winter. They had started roosting on October 21 ('94), and were using it in numbers from November 4 ('94) to March 24 ('95), but I only saw a few on March 31. A nest with 3 fresh eggs on March 13 ('88,

Wholey), and another with 1 fresh egg on May 13 ('82), are extreme dates. Young were still in a nest on June 5 ('92) and a family was still holding together on August 13 ('93). Sets are 3 of 2, 3 of 3, 18 of 4, 22 of 5, 1 of 7, and 1 of 10. On Dan's Mountain I only saw 5 or 6 crows from June 5 to 14 ('95).

Corvus ossifragus (490). Fish Crow.

Resident in tidewater Maryland. Five eggs nearly incubated were taken at Gunpowder on May 21 ('93, Fisher), 5 in the same condition on May 26 ('92), and a fresh egg, 5 days later, were collected in Queen Anne County. On June 8 ('94), 5 eggs, about to hatch, also 1 fresh egg were, noted at Ocean City. This last nest was only 8 feet from the ground in an alder bush (*Alnus maritima*), the others were all in the tops of high trees.

Family ICTERIDÆ—Blackbirds, Orioles, etc.

Dolichonyx oryzivorus (494). Bobolink.

Common migrant from April 28 ('95) to May 26 ('95), and from August 9 ('94) to November 8 ('88, Resler). Usually observed in the uplands in spring, they are also numerous there in the fall, while the marshes are alive with them. In April, '95, seven were shot from a flock at Cumberland by Z. Laney.

Molothrus ater (495). Cowbird.

Resident; wintering in tidewater Maryland, they are seldom seen in the uplands until spring (March 12, '92, Gray), when they become numerous, but the majority soon go north, and during summer their presence is chiefly shown by the alien egg in the small bird's nest. In fall, however, flocks of any number up to 300 or 400 are frequently seen between September 16 ('94) and November 19 ('94). On November 11, '94, I came across an enormous flock of birds, 10,000, or more; they entirely covered a 25-acre field. Careful investigation with a field glass showed about equal numbers of Cowbirds and Red-winged Blackbirds.

Extreme nesting dates are May 21 ('93), when 2 eggs were in a Robin's nest with 3 of the owners, and August 25 ('95), when a young bird was being fed by Indigo Birds. As a rule, only one egg is found in a nest in Maryland; the only instance of more, beside the one mentioned above, that I know of, is 3 in a Wood Thrush's nest with one of the owners, found in Talbot County on July 25, '95 (Tylor).

Xanthocephalus xanthocephalus (497). Yellow-headed Blackbird.

Near Baltimore specimens of this western bird have been taken as follows: On September 10, '91 a male was shot from a flock of Red-winged Blackbirds feeding in a wild oat marsh near Curtis Bay by Mr. Otto Nickel, who presented it in the flesh to A. Resler, in whose collection it now is. On September 18, '93, a female was shot from a bunch of Blackbirds at Patapsco marsh by Richard Cantler; this I secured for the Maryland Academy of Sciences. On October 1, '94, Mr. Edwin Schenck got another female at Patapsco Marsh; it was flying by itself. This one had the feathers removed by an over-active cook.

"A female was brought me on August 29, '92, that was killed from a flock of Blackbirds on the marshes adjoining Washington. This is the first record for the District of Columbia" (E. M. Hasbrouck, Auk, x, 92).

Agelaius phœniceus (498). Red-winged Blackbird.

Resident in tidewater Maryland, but not found in the uplands during winter. Nesting among the reeds, grass or bushes of swamps or wet meadows, I was rather surprised when a nest was found in a wild aster on top of a hill and fully a mile away from the nearest piece of swamp and one-eighth of a mile from the run. On May 13 ('94, Tylor) eggs were found in Talbot County; near Baltimore, from May 18 ('90) to July 26 ('81). Sets are 1 of 1, 10 of 2, 44 of 3, 30 of 4, and 3 of 5. At Vale Summit I only noted a single female on June 13, '95.

Sturnella magna (501).   Meadowlark.

Common resident; numerous in flocks in spring and fall.   On May 10 ('91, Wholey) 2 fresh eggs were found and on July 21 ('95) 5 fresh.   Sets are 1 of 4, 5 of 5, and 1 of 6.

Only a few at Vale Summit, June 5 to 14, '95.

Icterus spurius (506).   Orchard Oriole.

Common summer resident.   First noted at Washington on April 25 (Richmond); at Hagerstown, April 30 ('79, Small), at Knoxville, May 4 ('94, Fisher), and at Baltimore, May 5 ('95), while on May 7 ('93), they were common.   They leave as soon as the young are able to go with them, and but few individuals are seen after August 4 ('95); the last on September 8 ('95).

Eggs are noted from May 30 ('93), to July 12 ('83).   Sets are 2 of 3, 2 of 4, and 3 of 5.

Only one pair was at Vale Summit, June 5 to 14, '95.

Icterus galbula (507).   Baltimore Oriole.

Locally common during summer.   On April 29 ('93, Gray) both males and females were observed, but that they arrive earlier is shown by a pair having started to build on May 2 ('91). They are numerously recorded until September 8 ('95), a few until October 6 ('94).

At Waverly a pair started to build on May 2 ('91), and finished the nest on May 14.   The year before they did not start until the 15th, and the nest was finished on the 21st; both were placed quite close together and within 20 feet of the veranda.   On June 7 ('82) a nest contained 2 fresh eggs; on June 23 ('95) young were in a nest, while flying birds were still being fed on July 21 ('95).

At Hagerstown they were noted on April 27, '80, on May 12 they were building, and on the 18th they had eggs (Small). At Vale Summit, June 5 to 14, '95, they were numerous, the nests containing either eggs or small young.

Scolecophagus carolinus (509). Rusty Blackbird.

Very common during migration ; quite a number winter in tidewater Maryland. First noted in a large flock on September 30 ('94), the last were taken May 7 ('88, Resler). As a rule, they unite with flocks of Cowbirds, Red-winged Blackbirds or Purple Grackles, but sometimes they are seen in flocks by themselves.

Quiscalus quiscula (511). Purple Grackle.

Resident ; it takes remarkably severe weather to drive this species from tidewater Maryland in winter. In the uplands they are numerous from March 1 ('95) to July 20 ('95) and not uncommon until early in November. Nests with eggs are recorded from April 28 ('94, Tylor) in Talbot County and from May 6 ('92) in Baltimore County, until June 8 ('84). Sets are 2 of 2, 5 of 3, 5 of 4, 13 of 8, and 4 of 6. Nesting anywhere, a bunch of cedars seems to be preferred. The nest is usually placed in a crotch, but other situations are frequently chosen, notably the spaces in the rough sides of the Fish Hawk's nest. On May 14, '81, I found one in a hollow of a tree not seven feet from the ground, and on May 22, '92, another in a partially torn out Flicker hole. In '91-2-3, they nested inside the barn on the Hampden property, in Dulaney's Valley, placing their nests on the rafters and roof-sill (J. H. Pleasants, Jr.,).

Quiscalus quiscula æneus (511b). Bronzed Grackle.

"Occasionally east of the Alleghanies, from Virginia northward" (Manual, 380). " Mr. Ridgway notes for us 'several specimens seen' but considers it 'rare.' Mr. Wm. Palmer notes one in his possession shot by Lewis McCormick at Fall's Church, Va., and several obtained by Henry Marshall, at Laurel, Md.," (A. C., 73). One was shot in Prince George's County, by Fred. Zeller, (Smith. Report, '86, 665), and another at Laurel, Md., by George Marshall (*Ibid.* 724).

" Little more than a straggler, and individuals passing through here hardly remain to breed. However, I have a male shot April 6, '86, a date when ordinary *quiscula* is nesting, and a female shot April 2, '87, about the time ordinary *quiscala* are laying their first eggs " (C. W. Richmond, Auk, v, 19).

I have a male taken on March 1, '95, in Dulaney's Valley, and careful investigation, at short range with a field glass on March 10, 17, 24 and 31, showed that about $\frac{1}{3}$ of each flock of Grackle's were referable to this race. On April 21, '95, at least one male was observed.

Quiscalus major (513).   Boat-tailed Grackle.

This species nests in the stunted pines and alder bushes (*Alnus maritima*) that grow on our ocean beach, where nests have been found by Mr. C. W. Dirickson, of Berlin, and on July 7, '92, Mr. W. N. Wholey found a colony near Ocean City, the nests containing young of different ages; a deserted one, however, held 3 rotten eggs. In June, '94, I visited this locality and found the bushes burned down ; between Delaware and Virginia I did not see a single grackle.

On May 6, '93, one was with a flock of Purple Grackles at Townsend Street just west of Fulton Avenue (Gray), and on May 17, '93, one was noted at North Point.

A small colony was found at Smith's Island, Va., where they were breeding ; eggs and small young being found, May 16–24, '94 (Richmond).

Family FRINGILLIDÆ—Finches, Sparrows, etc.

Pinicola enucleator (515).   Pine Grosbeak.

⁜ I can find no reference to this species, except that Dr. Coues says : "Northern border of United States in winter, sometimes south to Maryland" (Key, 343). "An extremely rare and probably only accidental visitor in severe winters " (A. C., 56). "Occasionally to Maryland" (Birds N. W., 105). "We have no recent records, and the old ones are mere traditions—*i. e.*, we have no precise dates or records of specimens" (Richmond).

Carpodacus purpureus (517).   Purple Finch.

Common in flocks during winter, they were exceedingly numerous round Baltimore during the severe season of '92–3. Extreme dates are October 1 ('90, Resler) and May 31 ('93, Fisher).   At Washington it is given as "common in migrations, less so in mid-winter.   One was shot September 17 ('87), by H. W. Henshaw.   I saw a flock at Great Falls, Md., on September 24–5 ('89), and one was shot on May 13 ('85), by Dr. H. M. Smith" (Richmond).

"Not known at Hagerstown until May 3, '83, when they swarmed in the town" (Small).

Passer domesticus.   English Sparrow.

The first English Sparrows brought to this country were 8 pairs, liberated in Brooklyn in the spring of '51, but nothing is known of what became of them.   Two years later, 100 were liberated there and the importation was kept up for 30 years, the birds being liberated at widely different points of the country, 2500 or more birds being introduced.

"In June, '74, a few birds were brought to Baltimore by an English Captain and liberated in Franklin Square.   These are the first birds we have any knowledge of that came to Baltimore direct" (*Baltimore American*, August 31, '93).   At the following points in Maryland they introduced themselves,[1] and were noted in the following order:

'65 Hancock.	'77* Mechanicstown.
'68 Cumberland. ,	'77* Union Bridge.
'70 Williamsport.	'78* Frostburg.
'72* Manchester.	'78* Lonaconing.
'75* Boonsboro.	'78 Middletown.
'75* Oakland.	'78* Sharpsburg.
'76* Clearspring.	'79* Burkettsville.
'76* Hagerstown.	'79* Emmettsburg.
'76* New Windsor.	'79* Sandy Springs.
'76* Smithsburg.	'80 Grantsville.
'76* Taneytown.	'80 Salisbury.
'76* Westminster.	* About.

[1] For a full account of the introduction, etc., of the English Sparrow, see "The English Sparrow in North America," by W. B. Barrows, United States Department of Agriculture, Division of Economic Ornithology and Mammalogy, Bulletin No. 1, 1889.

Now they are all over the State and here to stay, being common in the cities and towns, while there is scarcely a house in the country without more or less of them round it. In Baltimore City they are frequently seen building in the winter; these, I believe, are only roosting nests. On May 17 ('91) eggs about one-half incubated were taken, and on August 16 ('91) birds, just hatched, were found. Sets are 2 of 2, 2 of 3, 9 of 4, 7 of 5, and 2 of 6.

Loxia curvirostra minor (521).   American Crossbill.

Very irregular in its movements; I have only observed this species once. On November 11, '94, about 25 were feeding on the cones of a small scrub pine, at the north end of Dulaney's Valley; they were quite wild, and did not remain ten minutes; when flushed, they flew clear out of sight.

At Lawyer's Hill, Howard County, about a half mile from Relay, in March, '90, Mr. C. Gamble Lowndes found "a flock of about 40 in a small bunch of pines; firing into them, two came to the ground dead, and several others, also dead, hung from the branches by their bills or claws, so that they were dislodged with difficulty. The others were all killed during a week with a small rifle, and were quite good eating. At night they roosted in a small ravine filled with pines and scrub oaks, and they spent the entire day in the clump of pines."

On January 15, '91, one was taken at Bush River, by Mr. Basil Sollers. Late in October, '89, two were shot out of a flock of about 20 in Talbot County (Tylor). At Cumberland "it is seldom seen; some years ago I killed one out of a flock with a cane" (Z. Laney).

" On May 23, '84, Mr. Geo. Marshall shot two, male and female, from a flock of five, near Laurel, Md. The female showed unmistakable evidence of having recently incubated. Two days later another male was shot in the same locality" (Smith. Report, '84, 146).

" On May 17, '85, an adult male and a young bird in the striped feather, barely able to fly, were seen by me in a pine sapling, a short distance beyond the city (Washington) limits" (Hugh M. Smith, Auk, ii, 379).

"For a long time regarded as extra rare and irregular in winter; in December, '87, they were numerous at Washington, and every local collector secured a series. Individuals were caught alive in the Smithsonian grounds and in the Agricultural Department Park ; they were seen until April 19, '88. In the following winter they were rather rare. In '90, again they were uncommon, but remained until late, individuals and flocks being noted on May 10, 16, 17, 21 and 24. Specimens taken on some of these dates showed no signs of breeding. I have few records for '91, '92 or '93, except that Mr. Ridgway, who lives at Brookland, D. C., near an extensive patch of pines, observed them all summer in small numbers, and still reports them up to June, '95. In the past winter, '94–5, Mr. Figgins reported them common near Kensington, Md., flocks appearing at short intervals throughout the winter. He mentions seeing hundreds on March 24; his last date is April 7, when about a dozen were seen. Mr. R. S. Matthews and I saw two small flocks late in April '95 " (Richmond).

Loxia leucoptera (522). White-winged Crossbill.

"South in winter to or beyond 40° " (Manual, 393). Audubon mentions having secured a specimen "in Maryland a few miles from Baltimore " (iii, 191), and Mr. Henry Marshall has one, shot at Laurel about '74 (A. C., 57).

Acanthis linaria (528). Redpoll.

"Rare and irregular, perhaps only an occasional visitant in severe winters " (A. C., 57). One specimen, a female, was taken by Dr. T. H. Bean at Fort Runyon, Va., on February 19, '75 " (Wm. Palmer, Auk, xi, 333). Audubon says (iii, 121): "I have seen several that were obtained near Baltimore, Md."

Spinus tristis (529). American Goldfinch.

Common resident; they commence flocking by the middle of September and hold together until early in June. On April

30 ('93), however, I saw an early pair mating, the male being
in nearly full summer plumage. While no doubt they nest
earlier, the first note I have is of a pair building on July 12
('91), the latest date is September 9 ('94), four fresh eggs.
Sets are 1 of 3, 3 of 4, 2 of 5, and 1 of 6. At Vale Summit
they were not numerous and still in flocks June 5 to 14, '95.

Spinus pinus (533).   Pine Siskin.

Irregular winter visitor, usually found among the pines.
On November 24, '92, quite a large flock was near St. Mary's
Industrial School; on January 29, '93, a flock of about 20 was
at St. George's Avenue, and on March 1, '79 (Resler), a number
were at Bayview.

At Washington, "very abundant some winters between October and April; at times, several years will pass without one being
seen" (Richmond).

Plectrophenax nivalis (534).   Snowflake.

Irregular winter visitant in cold seasons. On February 6, '92
(Gray), a flock of eight was seen at Calverton; on February
10, '95, two were noted on the Falls Turnpike near Melvale,
and on the same day (Fisher) two were seen on the drive round
Druid Hill Park lake.

One specimen has been taken at Washington (Richmond).

Calcarius lapponicus (536).   Lapland Longspur.

"South in winter to northern United States, sometimes (rarely)
as far as South Carolina" (Manual, 404). On February 10,
'95, just after the blizzard, all our roads being impassable to
vehicles, I walked out to Lake Roland. At the point where
the Northern Central Railroad crosses the Falls Turnpike a
small part of the bank had been cleared of snow by the wind,
on this were about 20 Longspurs, with 8 or 10 Song Sparrows
and about 25 Tree Sparrows. They allowed me to approach
within three feet.

"On December 11, '86, while Dr. Fisher and I were riding along the road to Falls Church, Va., and distant from Washington perhaps four miles, we saw a flock of 15 or 20 Horned Larks by the roadside. Scattered through the flock were half a dozen or more Longspurs, one of which was secured. Comparatively little collecting has ever been done about Washington in winter, and to this fact, rather more than to its excessive rarity, is due, I am persuaded, the absence of the species from the local lists. Although probably not a regular migrant, the species occurs here in small numbers, I am inclined to believe, every hard winter. However, it is to be remarked that the records for this bird so far south are very few " (H. W. Henshaw, Auk, iv, 347).

Poocætes gramineus (540).   Vesper Sparrow.

Resident; abundant during migrations from March 20 ('92,) to May 23 ('93, Wholey), and from September 22 ('95) to December 2 ('94). During summer they are not very common, and only a few winter with us. On June 19 ('90, J. H. Fisher, Jr.) 3 slightly incubated eggs were taken. At Hagerstown it is given as "not common; occasionally seen in June, July, September and October" (Small). At Vale Summit on June 11, '95, I found one pair feeding young just out of the nest.

Ammodramus princeps (541).   Ipswich Sparrow.

"Breeding on Sable Island, N. S.; in winter migrating along the Atlantic coast south to Virginia." (Manual 407).   "Winter resident along the New Jersey coast; not abundant, though probably regular."   9 specimens are recorded between November 16 ('80) and April 3 ('89) (Birds, E. Pa. and N. J., 112–3).

"Those who care to visit in winter the bleak, wind-swept sand hillocks of our Atlantic coast will find this bird much less rare than it was once supposed to be " (Chapman 291).

Ammodramus sandwichensis savanna (542a). Savanna
Sparrow.

Fairly common migrant, some possibly wintering; noted from
October 6 ('95,) to November 9 ('92, Resler), and from
March 18 ('91, Resler) to May 12 ('92, Pleasants). At Wash-
ington "4 were noted on October 14, '94, (J. D. Figgins and
Wm. Palmer); a few winter" (Richmond). At Hagerstown
"quite common during the fall migration of '80" (Small).
No doubt it winters along our sea-coast for "at Cape May
and probably all along the coast of southern New Jersey the
Savanna Sparrow is an abundant winter resident" (Birds, E.
Pa. and N. J., 113).

Ammodramus savannarum passerinus (546). Grasshop-
per Sparrow.

A few wintering; this species is common from April 10 ('95,
Fisher) until November 4 ('94). It is usually flushed from the
ground, but during the breeding season it mounts on a weed or
even to the top of a fence to sing the very peculiar song from
which it derives its name.

Although very numerous in the breeding season, the nest is
seldom found, it being placed on the level ground in the open
field. Young out of the nest were being fed on June 4 ('93),
and on August 11 ('94) birds 4 or 5 days old were in a nest.
Sets are 3 of 3, and 3 of 4.

Common at Washington from March 30 ('95) to October 21
('87, Richmond). At Vale Summit a number were observed
June 5 to 14, '95.

Ammodramus henslowii (547). Henslow's Sparrow.

"Known from this region for many years, this has been considered one of the rarest sparrows, although always found in
one or two localities. On May 30, '92, while exploring Johnson's Gully, Maryland, 16 miles south of Washington, a large
colony was found" (E. M. Hasbrouck, Auk, x, 92). "Not

uncommon in several localities near Washington. At Kensington, Maryland, where it is rather common, Wm. Palmer found a nest with young on June 1, '94, and J. D. Figgins shot 2 on October 14, '94. It is also rather common at Laurel, where, on April 10, '89, Mr. Ridgway shot a male " (Richmond) and Geo. Marshall secured specimens on May 4 and 21, and August 11, '94. In Howard County, specimens were taken on April 17, and August 3, '93 (Resler).

Ammodramus caudacutus (549). Sharp-tailed Sparrow.

" Abundant summer resident on the salt marshes along the New Jersey coast and for some distance up the shores of Delaware Bay. Mr. I. N. DeHaven finds a few 'sharp-tails' nearly every winter on the Atlantic City meadows, but they are by no means common at this season " (Birds, E. Pa. and N. J., 114).

On June 7, '94, I found a number along the beach some miles south of Ocean City, Maryland.

Ammodramus caudacutus nelsoni (549a). Nelson's Sharp-tailed Sparrow.

" Fresh water marshes of the Eastern United States, and during migrations, to marshes of the Atlantic coast, Massachusetts to South Carolina" (Manual, 413). " Rare transient on the New Jersey coast, though probably of regular occurrence. The least common of the three races of Sharp-tailed Sparrows on our coast, it will probably be found to be more abundant in fall than in spring. Specimens have been secured May 9 and October 2, '92, by Mr. I. N. DeHaven " (Birds, E. Pa. and N. J., 115).

In Virginia the following specimens have been taken, "one by C. Drexler in September, '62, another by E. J. Brown, at Cobb's Island on May 11, '92, and a third by myself on 4 mile Run marsh, Alexandria County, on September 18, '93," (Wm. Palmer, Auk, xi, 333).

Ammodramus caudacutus subvirgatus (549b). Acadian
Sharp-tailed Sparrow.

" Was found associated with the other two races at Atlantic
City on October 2, '92, and a number of specimens shot. It
appeared to be more numerous than *nelsoni* but less so than true
caudacutus. This species apparently winters further south, for
all the winter specimens so far taken by us in New Jersey were
caudacutus. Doubtless more careful search will show the *nelsoni*
and *subvirgatus* to be of regular occurence in both migrations"
(Witmer Stone, Auk, x, 85)." Regular transcient visitor on the
New Jersey coast marshes" (Birds, E. Pa. and N. J., 114).

Ammodramus maritimus (550). Seaside Sparrow.

"Abundant summer resident on the New Jersey coast marshes
and on the shores of Delaware Bay, arrives at Atlantic City
about April 20 and departs October 15. In Cape May County,
Mr. W. L. Baily secured several specimens February 22, '92,
which would indicate that a few of these birds winter there occasionally" (Birds, E. Pa. and N. J., 115). Between, May 14
and 28, '94, at Smith's Island, . Va., they were "breeding and
quite numerous, we secured 43 specimens of this bird." (E. J.
Brown, Auk, xi, 259). " On Smith's Island I found 3 · nests
May 18, '94, with 4, 5, and 5 eggs and on May 23, '94, another
with 4 eggs " (Richmond).
On May 27, '93, I came accross a pair of these birds, one of
which I secured, but careful search failed to find the nest. They
were in a marsh on Start's Point, between Chester River and
Langford's Bay, in Kent County. On June 6, '94, I found a
nest with 5 eggs nearly hatched, on our ocean front quite close
to the Virginia line, where they were rather common.

Chondestes grammacus (552). Lark Sparrow.

This bird of the Mississippi Valley occasionally strays to the
Atlantic coast. "Up to date, our knowledge of the occurrence
of the Lark Finch in the neighborhood of Washington is limited

to the capture of a single specimen by Mr. Roberts, on August 25, '77 (at Fort Runyon, on the Viginia side of the Long Bridge, A. C., 66), and the observation of 2 individuals in the Smithsonian grounds by Mr. Ridgway (on August 27, '77, Auk, iii, 43). To the above is to be added the capture of a second specimen, an adult male, on August 8, '86, by the writer" (H. W. Henshaw, Auk, iii, 487).

Zonotrichia leucophrys (554).   White-crowned Sparrow.

Regular, though not common, migrant from October 7 ('94) to October 23 ('92), and from May 6 ('93 and '94, Wholey) to May 21 ('92, Gray), and on the same day at Sandy Springs (Stabler). At Washington, "from October 13 ('93), when several were seen by H. W. Henshaw and A. K. Fisher, to November 25 ('88), when one was shot by J. D. Figgins, and from the middle of April to May 17 ('86), when one was shot out of a flock of about a dozen" (Richmond).

Zonotrichia albicollis (558).   White-throated Sparrow.

Winter resident; common; more numerous during migrations. Noted from September 21 ('93, Gray), and common from September 29 ('95) to May 13 ('92), the latest record being May 19 ('95), when two, a male and female, were found in a brush pile in Dulaney's Valley, and some years ago one was taken on the same date by Mr. A. Resler. At Washington "one was shot on September 15 ('89), and they were common from September 30 ('90, '94) until May 20 ('88), when one or two were noted and a male was shot by W. F. Roberts. On May 21 ('86) one was shot and ('88) several were heard singing " (Richmond).

Spizella monticola (559).   Tree Sparrow.

Equally numerous during winter with the " Snowbird, " this species is not so commonly observed, as it does not come round the house, preferring the briar and weed patches.   On October 7 ('94) I found it common in flocks with Snowbirds; and on

April 29 ('94) quite a number were in a flock with Field and Chipping Sparrows. At Washington "from late in October to early in April" (Richmond).

Spizella socialis (560).   Chipping Sparrow.

First noted on March 16 ('94), and common from March 25 ('94) to November 27 ('92), 8 or 10 were seen on December 2 ('94), and about a dozen on December 4 ('92). At Washington, "some spend the winter, but not in numbers sufficient to be detected every season. On March 12 ('90, J. D. Figgins) four were seen. Common from the end of March to November 24 ('89), one was seen on December 14 ('90)" (Richmond). Mating on April 22 ('94), the first nest with eggs was found on May 10 ('91), the last on August 28 ('92), and on September 16 ('94) young just out of the nest were observed. Sets are 2 of 2, 27 of 3, and 14 of 4.

At Hagerstown their arrival is noted on March 30, '79, March 18, '80, and March 21, '81. Under date of April 15, '83, is the following: "Has any one ever noted the Chippy's fondness for the sap of the grape vine? They make a regular habit of drinking large quantities when the vines have been trimmed in spring. Do they do this for water, or for any nourishment there is in the sap? I rather think the latter, as they drink too much for simply quenching their thirst" (Small).

Only a few at Vale Summit, June 5 to 14, '95.

Spizella pusilla (563).   Field Sparrow.

This is our most common breeding sparrow, being exceedingly numerous from March to November. During mild winters, a number stay with us, and even during the cold winter of '92–3 a flock was seen on December 4, about 20 on January 4, and a single bird on January 22. At Washington it is noted as "a common regular winter as well as a summer resident" (Richmond). Nests with eggs are noted from May 8 ('91) to August 16 ('91). Sets are 5 of 2, 22 of 3, and 15 of 4.

At Vale Summit they were not common, June 5 to 14, '95.

Junco hyemalis (567).   Slate-colored Junco.

A winter resident; first noted on September 28 ('93, Gray);
it was common on September 30 ('94), but, as a rule, they do not
become numerous until the middle of October, remaining so
until April 23 ('93, Fisher), the last is recorded on May 1 ('74,
Resler. '92). At Washington "from September 30 ('94),
when several were seen by Messrs. Figgins Matthews and Palmer,
to April 27 ('90), when several were seen. On May 1 ('87),
one was seen and on May 4 ('—) one was shot" (Richmond).

Junco hyemalis shufeldti (567b).   Shufeldt's Junco.

A western bird. "On April 28, '90, my son, A. W. Ridgway, shot a female of this sub-species near Laurel, Maryland.
It was shot out of a small flock in which, my son thinks, were
others of the same kind, but he may have been mistaken"
(Robt. Ridgway, Auk, vii, 289).

Melospiza fasciata (581).   Song Sparrow.

Common resident; liable to be heard singing at any time,
even in the depth of our most severe winters. Nesting dates
range from May 3 ('91, Wholey), 5 fresh eggs, to September
11 ('92), when young leave a nest. Sets are 1 of 2, 5 of 3, 13
of 4, and 8 of 5. Situations of nests noted show 8 built up
from the ground in bushes, etc., 7 on level ground in grass, and
8 in hollows in the sides of banks, while one was in the side of
a haystack (Fisher).

At Hagerstown, Small says, "one pair has lived in our yard
and built two nests each year; averaging from 8 to 10 feet
above the ground; the lowest was 4 feet up, and one was fully
25 feet up in a western vine growing against the house. Cow-
bird eggs were twice deposited, when all the eggs were thrown
out and they started to build a new nest the next day. Twice,
one of this pair was killed, but a few days later the survivor
secured a mate."

Only a few were seen at Vale Summit, June 5 to 14, '95.

Melospiza lincolni (583),   Lincoln's Sparrow.

" In the Atlantic states it is apparently rare ; in the course of all my collecting I never saw it" (Birds, N. W., 136). L. M. McCormick mentions 2 specimens obtained in the District of Columbia, without date (Auk, i, 397). " Mr. Henshaw collected on the Virginia side of the Potomac 3 specimens in May, '85. Wm. Palmer has taken 2 birds ; and Mr. Ridgway has noted this species on several occasions near Laurel, Md. (Auk, v, 148). "On May 7, '92, J. D. Figgins got one at Kensington, Md., on May 12, '87, R. Ridgway got one at Gainesville, Va., and on May 18, '84, H. W. Henshaw got one, the first on record for the District of Columbia, Wm. Palmer shot one on September 30, '94, and Robert Ridgway shot one on October 18, '89, at Laurel" (Richmond).

Melospiza georgiana (584).   Swamp Sparrow.

Occasionally noted during winter, and common from September 30 ('94) to November 11 ('94), and again from April 8 ('93, Gray) to April 29 ('94) ; extreme dates are September 16 ('94) and May 12 ('94, Wholey).   At Washington, "from September 28 ('90) to the last of October, when they sometimes swarm in suitable places; a few winter, and in April they again become common, the last being noted on May 16 ('88) " (Richmond).

" Resident ; though much more abundant during the migrations than at other times.   The Swamp Sparrow breeds on the marshes of Tinicum Township, Delaware County, Pa., and in other similar situations, but seems to be rather locally distributed during the breeding season in southern Pennsylvania " (Birds, E. Pa. and N. J., 118).

Passerella iliaca (585).   Fox Sparrow.

Common during migrations, from October 21 ('94) to December 16 ('94), and from February 13 ('92, Gray) to April 23 ('92, Gray), a few occasionally wintering.   On November

5, '93, I saw between 400 and 500 in Dulaney's Valley, one of which gave me a wild burst of song. At Washington, recorded "from October 18 ('94) to April 15 ('94). Two were shot on April 21 at Laurel, by Geo. Marshall" (Richmond); noted at Hagerstown in May, '80, and also "a few all winter" (Small).

Pipilo erythrophthalmus (587). Towhee.

Common summer resident, occasionally wintering. Noted from March 8 ('94) to November 13 ('92), and common from April 2 ('93) to October 21 ('92); nests with eggs range from May 13 ('94), to August 28 ('91, Fisher). Sets are 8 of 3, and 3 of 4. Six nests were on the ground, the others elevated, the highest being 6½ feet up in a cedar.

During the severe winter of '92–3, Mr. W. N. Wholey and I watched a flock from December 8 to February 14, seeing it 3 or 4 times a week. Neither of us could ever count the flock; sometimes there seemed more, sometimes less, but we estimated it to average about 50 birds. There was snow on the ground all this time and the temperature went down as low as 1°, and for over a month the maximum kept below 32° (*Fahrenheit*).

In Somerset County they were numerous from November 15 to 21 ('94, Fisher).

At Vale Summit they were common, June 5 to 14 ('95).

Cardinalis cardinalis (593). Cardinal.

Resident, old birds generally seen in pairs, the young (presumably) going in flocks in fall and winter, occasionally a few pairs of adults may be found quite near each other. Nesting dates range from May 3 ('92, Wholey), a nest ready for eggs, to August 18 ('83), 3 eggs. Sets are 1 of 2, 17 of 3, and 1 of 4.

Habia ludoviciana (595). Rose-breasted Grosbeak.

A migrant, seen at irregular intervals and not to be procured every year; it has been fairly well noted from April 28 ('93, Resler) to May 20 ('76, Resler) and from September 11 ('80, Resler) to September 25 ('93, Gray).

"A regular migrant at Washington, not rare, but of irregular abundance from May 3 ('91 and '92) to May 16 ('88 and '90) and from August 29 ('87), when Mr. R. Ridgway shot one at Gainesville, Va., to September 28 ('89)" (Richmond).

On Dan's Mountain, in Alleghany County, I found 2 pairs, one located at Pompey Smash, the other about a mile distant at Lauertown. Morning and evening, from June 5 to 14, '95, both the males could be heard singing. On June 8, I found the nest of the Pompey Smash pair; it contained a young bird not 24 hours old and 2 infertile eggs. On the 14th I again visited this nest; on being disturbed the young one scrambled out of the nest and fell to the ground; I replaced it and hope it arrived at maturity, and will live to a green old age.

Guiraca cærulea (597).   Blue Grosbeak.

This large sized edition of the Indigo bird seems to nest more or less regularly in southern Maryland, Washington being about its northern limit, where it is noted as "rather rare, from first week of May to middle of September" (A. C., 68). "During the summer of '87 a pair nested twice on my father's farm, about a half-mile east of the District of Columbia. On June 24 I took the first nest and 4 eggs from the fork of a peach tree, about 7 feet up. . . . In August evidently the same birds nested in a small cedar. I have observed them every season since, but have found no more nests" (A. B. Farnham, Oologist, viii, 219–20).

At Washington "a male was seen on June 30, '89; a pair, male and female, were taken by Wm. Palmer on August 15, '92, and several were seen and one shot on September 19, '86. At Kensington, Md., a nest was found early in June '95, by J. D. Figgins, and another with 4 eggs was found at Laurel about June 10, '95, by Geo. Marshall " (Richmond). One was taken in Howard County, on July 29, '93 (Resler). On August 3, '91, a pair were seen feeding flying young on the Windsor Road, about a half-mile this side of Powhattan (Gray and Blogg). At Hagerstown "a male with testes as large as peas was shot by A. J. Jones on May 23, '82 " (Small).

Passerina cyanea (598). Indigo Bunting.

Common summer resident; first noted on April 30 ('92, Gray) and common from May 2 ('93) to September 30 ('94); the last was taken on October 5 ('87, Resler). At Washington, "noted on April 29 ('94, Currie, Preble and Hasbrouck); it was still common on September 30 ('90, '94), while one was found on October 15 ('90); and on December 13 ('87) Mr. M. M. Green shot one, it was 'fat and healthy'" (Richmond).

A nest ready for eggs on May 25 ('90) contained 3 on June 1, and on August 29 ('91); a brood of young left a nest on being disturbed. Sets are 1 of 1, 5 of 2, 17 of 3 and 5 of 4.

At Vale Summit they were common, June 5 to 14, '95.

Spiza americana (604). Dickcissel.

"A summer resident. This bird used to arrive regularly about the first of May, and leave towards the end of September, meanwhile being very abundant. . . . Now, however, the birds seem to have forsaken us, few, if any, having been heard for the past few years" (A. C., 67). "Extremely rare; this bird is said to have been abundant formerly, but it appears to have withdrawn almost entirely from this vicinity. A male, seen by Mr. Henshaw about the last of May, '87, was very likely nesting " (C. W. Richmond, Auk, v, 22). "At Jefferson, Maryland, Mr. J. D. Figgins got one on August 4, '90, and 2 more the next day" (Richmond).

The case seems to be paralleled near Baltimore. I have been several times told of its former abundance, and how easily it was taken with bird-lime. The only late item, however, is "on May 7, '92, near Wyndhurst Avenue, on the Baltimore and Lehigh Railroad track, I heard a strange bird song; locating it, I saw a male Black-throated Bunting, it was in plain sight and identification is positive" (Wholey). Specimens were taken by Mr. Resler on May 20, '76, June 10, '76 and October 2,

'80, at Back River.   At that time they were quite common, and no special attention was paid to them.

Family TANAGRIDÆ—Tanagers.

Piranga erythromelas (608).   Scarlet Tanager.

Summer resident; generally dispersed, but local and not very common. Noted from May 4 ('92, Resler) to October 7 ('93, Pleasants). At Washington "from April 25 ('95) to October 7 ('88, W. T. Roberts)" (Richmond).

Nests with eggs are noted from June 2 ('88, J. H. Fisher, Jr.) to July 18 ('93, Pleasants), while in Howard County three fresh eggs were collected on August 1 ('92, Resler). Sets are 2 of 3 and 3 of 4.

Only a single pair at Vale Summit, June 8, '95.

Piranga rubra (610).   Summer Tanager.

Summer resident ; about as numerous as the last species, noted at Washington from April 28 ('94, Richmond) and at Baltimore from April 30 ('93, Blogg) to September 12 ('93, Gray). Eggs are recorded from June 7 ('93, Wholey) to July 10 ('93, Bloggs). Sets are 4 of 3 and 1 of 4.

In '90–1–2, Scarlet Tanagers occupied a piece of open woods back of our house in Dulaney's Valley; in '93, there were no Scarlet Tanagers, but several pairs of Summer Tanagers; in '94, no Tanagers appeared; but this year ('95), four or five pairs of both species spent the summer, some of them nesting quite close to the house.   Both species have been more numerous this year than I have known them before.

Family HIRUNDINIDÆ—Swallows.

Progne subis (611).   Purple Martin.

Common summer resident.   First noted in Kent County on March 30 ('95, Fisher), and in Baltimore County from April 2 ('93) to October 15 ('93).   In spring they are generally first observed at the bird boxes, these they leave as soon as the young are able to go with them, the latest date being August 27 ('94), when only one pair was left. On July 28 ('95), 6 were in a small dead tree in Dulaney's Valley; as there were only one adult male, this family presumably had started to migrate. In this same tree on August 12 ('94), at 10.40 A. M., I saw 18 (3 adult males) all sound asleep, with their heads under their wings. On being flushed by a knock on the tree they flew round for a few minutes and returning, settled themselves again, evidently played out from a long migration flight. At Bay Ridge, towards dusk, on August 18 ('95), quite a large number flew south in an uneven but regular column.

On May 18 ('92) a box held 8 nests in various stages of construction, from just started to ready for eggs, and on May 29 ('94) another had 21, ranging from just started to having the complete set of 5 eggs, while as late as July 9 ('95, Henninghouse) 3 fresh eggs were taken. Sets are 3 of 5. Usually nesting in the boxes put up for them, on June 12, '94, I found nests with young under the eaves of cottages at Ocean City. At Washington they formerly nested in the tops of the columns of the Treasury Building, now "in crevices of the Post Office Department Building and the Masonic Temple" (Richmond). At Bay Ridge, on June 28, '93, several pairs were nesting in the hoods of electric lamps (Gray). At Cumberland on June 4, '95, I found them in the electric lamps, and also in cornices of buildings.

Petrochelidon lunifrons (612).   Cliff Swallow.

Summer resident, extremely local and not common; but a colony may locate their nests under the eaves anywhere, occupy them any number of summers, then, without apparent cause, leave, sometimes coming back after one or more years. First seen on April 16 ('93), when 26 were flying in a loose flock; the last were also seen in a loose flock on September 3 ('93). At Washington, September 12 (A. C., 52). Nesting dates range from June 7 ('82), five fresh eggs, to July 23 ('93), when young birds were still in nests. On July 22 ('93), however,

I saw quite a number on the telegraph wires, young and old, ready to go south. Sets are 1 of 1, 1 of 2, 2 of 3, 1 of 4, 1 of 5.

Chelidon erythrogaster (613).   Barn Swallow.

This, our swallow proper, is an exceedingly numerous summer resident, there being scarcely a barn or out-building that has not its swallow nests on the rafters. First noted on April 3 ('80, Resler), on the 15th ('93) they were numerous, remaining so until September 20 ('90, Resler). Birds congregated on the telegraph wires, ready to migrate, have been seen as early as July 5 ('94). Nesting dates range from May 29 ('81), slightly incubated eggs, to August 9 ('93), small young birds. Sets are 1 of 1, 4 of 3, 6 of 4, and 10 of 5. At Washington they were noted March 30 ('90), and nests with 1 and 2 eggs were found on May 16 ('86, Richmond). At Hagerstown on March 20, '80 (Small).

A small colony was at Vale Summit June 5 to 14, '95.

Tachycineta bicolor (614).   Tree Swallow.

Sparingly distributed along the shores of tidewater Maryland in summer, this species may be seen anywhere during migrations, when it is abundant. Noted from March 30 ('92, Resler) to October 17 ('94, Resler); quite a large number were flying south in a loose, desultory manner over Ferry Bar on August 12 ('95). At Washington it is recorded from "March 30 ('90), to October 14 ('94), when there must have been a wave of them, as Wm. Palmer got one and saw several at Kensington, Md.; Geo. Marshall saw one at Laurel, Md., and R. Ridgway saw 2 or 3 at Gainesville, Va." (Richmond). On October 14, '94, there was quite a large number in Dulaney's Valley. Several were seen at Harper's Ferry, April 8, '87, by F. L. Washburn (Birds Vas., 77). At Hagerstown in September, '82, and May, '83 (Small). Nests with eggs are noted from May 12 ('94, Fisher) to July 3 ('93, Blogg). Sets are 1 of 4, 2 of 5, and 1 of 6.

Clivicola riparia (616).   Bank Swallow.

Common summer resident.   Quite a number were at Loch
Raven on April 14 ('95), but they were down the necks earlier,
for on April 19 ('95, Fisher), at Gunpowder, a new hole had
been dug about a foot deep and two others a few inches.   The last
birds were noted September 12 ('85, Resler).   At Washington,
from April 25 ('86, '94) to September 15 ('90, Richmond).
On July 17, '91, a large number, possibly between 3000 and
4000 were on a ducking blind at the mouth of Back River;
they kept up an incessant twitter and were evidently ready to
leave for their winter home.   May 12 ('91), slightly incubated
eggs, and July 17 ('92), young birds, are extreme nesting dates.
Sets are 3 of 2, 2 of 3, 3 of 4, 3 of 5, 2 of 6, and 2 of 7.

Stelgidopteryx serripennis (617).   Rough-winged Swallow.

Summer resident, but not as numerous as the Bank Swallow.
Observed at Washington April 8 ('92, R. Ridgway, Auk, ix,
307), at Baltimore, April 19 ('79, Resler); the last recorded at
Washington was shot on September 3 ('94, Richmond).   Sets of
eggs, noted from May 13 ('83) to June 15 ('84), are 2 of 4, 1
of 5, 1 of 6, and 1 of 7.   "Numbers of these birds breed along
the Potomac River in crevices of the rock; . . . a nest of
7 eggs found during June, '87, contained six eggs of this
species and one of the Barn Swallow" (C. W. Richmond, Auk,
v. 23).   At Hagerstown, Small says ('80–'81), "the Bank
and Rough-winged Swallows are about equally common."

Family AMPELIDÆ—Waxwings.
Ampelis cedrorum (619).   Cedar Waxwing.

Resident, roving in flocks all the year, from which, in their
leisurely manner, a pair will detach themselves and go to house-
keeping, generally in July or August.   Eggs are noted from
June 17 ('92, Blogg) to August 21 ('92).   At Vale Summit
a nest had 5 fresh eggs on June 11 ('95).   Sets are 1 of 3,
2 of 4, and 4 of 5.

Family LANIIDÆ—Shrikes.

Lanius borealis (621).   Northern Shrike.

Probably a regular, though rare winter visitor. On October 26, '87, one was taken on Patapsco Marsh (Resler); December 2, '94, one in Dulaney's Valley; January 10, '93, one at Bay View (A. Wolle); February 11, '92, one at Powhattan (Gray); February 19, '93, one was seen in Dulaney's Valley, and on February 25, '93, one was seen just west of the city (Gray). At Washington, "one was shot on December 26, '87; another on January 10, '91, and one on February 10, '46. Numbers of others have been taken" (Richmond).

Lanius Ludovicianus (622).   Loggerhead Shrike.

A common resident in the south Atlantic States, wandering north in winter.   Near Baltimore single birds are numerously recorded from August 26 ('92, Blogg) to April 23 ('93, Gray), and near Washington from "August 11 ('89) to April 6 (—) ; dozens have been reported, mostly, however, in midwinter" (Richmond). At Hagerstown "only seen a few times. One was seen in May eating a large beetle under the spruce in our yard, and one was noted in December, '80" (Small).

Family VIREONIDÆ—Vireos.

Vireo olivaceus (624).   Red-eyed Vireo.

Very common summer resident. First noted on April 23 ('93); a week later they were numerous everywhere, remaining so until October 6 ('94), the latest date being of a single bird on October 11 ('93).   At Washington "from April 22 ('91) to October 14 ('94, Palmer and Figgins), specimens were taken —one on October 17, '90; four on October 30, '94 (Palmer and Figgins), and one on November 11, '88 (Figgins)" (Richmond).

Nests with eggs range from June 8 ('85) to July 31 ('93), while young birds not long out of the nest and still being fed by the parents were seen on September 4 ('92).   Sets are 11 of 2, 29 of 3 and 3 of 4.

Vireo philadelphicus (626). Philadelphia Vireo.

"Not very common in the Atlantic States" (Key, 332). "A regular but rather rare spring and fall migrant, arriving here late in April or early in May; after the 20th of May it is seldom seen in Pennsylvania until it migrates southward in September" (Birds Pa., 264). "A very rare migrant;" 10 specimens are recorded between September 11 ('80) and October 6 ('91) (Birds E. Pa. and N. J., 126). "One taken on the Virginia side of the Potomac near Washington, on May 17, '88" (Wm. Palmer, Auk, vi, 74). "Wm. Palmer also got one or two additional ones, and J. D. Figgins shot one on September 16, '94, at Kensington, Md." (Richmond).

Vireo gilvus (627). Warbling Vireo.

Of local distribution, in some parts of Maryland it is a summer resident, but near Baltimore I have only found it as a migrant. On May 5, '82 (Resler), one was taken, the only spring note. In '94 I found it quite numerous in Druid Hill Park on August 21, 22 and 24, and again on October 4 and 6, but not between times. At Washington it is "a summer resident, not very common, from April 28 ('89) to September 10." Mr. Figgins found it very common at Kensington, Md., on August 23, '94 (Richmond).

At St. Michael's, Talbot County, on June 15, '94, a pair were very lively in a shade tree. At Hagerstown, "the Red-eye and Warbling Vireos are to be heard all through the long summer days; they both breed in town, but the Warbling is decidedly the most common, staying with us from the first week of May to October" (Small). In Cumberland, on June 4, '95, quite a number were in the shade trees.

Vireo flavifrons (628). Yellow-throated Vireo.

Regular, but not very common summer resident, more numerous during migrations. Noted from April 13 ('90, Pleasants) to October 6 ('94). In Dorchester County they were numerous,

and specimens were taken on April 6, '93 (R. C. Watters).
"A common summer resident at Washington, from April 20
to the middle of September. Wm. Palmer found a nest with
3 small young on June 1, '94" (Richmond). On May 27 ('93,
Blogg) a nest held 3 half-incubated eggs; on June 15 ('93,
Fisher) one contained young birds, and on June 19 ('92, Wholey)
one had 4 eggs ready to hatch.

Vireo solitarius (629).   Blue-headed Vireo.

A regular migrant, not common. Noted from April 19 ('93,
Resler and Wholey) to May 11 ('93, Pleasants), and from September 26 ('94, Resler) to October 22 ('92, Gray). During
spring single birds are usually seen; in fall, small parties of 4
or 5. At Washington "not uncommon in migrations, April 13
('88, Hasbrouck) to May 10 ('85) and October 8 ('88) to October
26 ('90)" (Richmond). At Hagerstown "a few were seen in
the spring migrations; they had a decided partiality for some
climbing roses, and were very unsuspicious" (Small).

Vireo noveboracensis (631).   White-eyed Vireo.

Common summer resident from April 22 ('91, Resler) to
September 23 ('94). At Washington "from April 12 ('90) to
October 12 ('90)" (Richmond). Nesting notes range from May
27 ('95, Fisher), 2 fresh eggs, to August 6 ('93), 2 birds about
two-thirds grown. Sets are 1 of 2, 3 of 3 and 1 of 4.

While the nest is usually placed near water, or in a swampy
place, on July 4, '94, I found a nest containing young nearly
ready to fly, on top of a high, dry hill, fully ⅛ of a mile from
the nearest spring.

Family MNIOTILTIDÆ—Wood Warblers.

Mniotilta varia (636).   Black and White Warbler.

First noted on April 15 ('93, Gray; '95, Resler), and numerous
on April 23 ('92, '93). Gradually thinning out as the breeding
season comes on, it is again abundant from early in August to

September 16 ('94), after which it is occasionally noted until
October 9 ('89, Resler).   At Washington " from April 8 ('88)
to October 18 ('92)" (Richmond).   The nest, being placed on
the ground, is seldom found ; those noted are: May 16 ('91,
Blogg), 3 eggs nearly hatched ; June 1 ('89, Blogg), 4 birds
and a rotten egg; July 4 ('92), 3 birds a few days old.

On Dan's Mountain they were very common.   On June 5,
95, I found a nest with 5 nearly fresh eggs ; it was placed in a
crevice, about 4 feet up the perpendicular face of Dan's
Rock, on the summit of the mountain.

Protonotaria citrea (637).   Prothonotary Warbler.

"Rare or casual on the Atlantic coast north of Georgia"
(Manual, 484).   One was seen on May 2, '61, near Washington
(A. C., 42.)   "On May 17, '88, near Laurel, Mr. P. L. Jouy
and I noticed a Prothonotary Warbler, and later, when we were
in company with Geo. Marshall, another (or possibly the same
one) was seen, but we failed to secure either" (R. Ridgway).
On August 25 '95, I saw 2 in Dulaney's Valley ; they were
fussing with one another and allowed of quite close approach.
"On May 11, '94, an adult male was taken at Mt. Vernon, Va.,
by E. M. Hasbrouck" (Richmond), and Captain Crumb has
taken one at Cobb's Island (letter to W. H. Fisher).

Helmitherus vermivorus (639).   Worm-eating Warbler.

Sparingly resident, from May 2 ('90, Gray) to September 10
('92, Resler).   At Washington "from April 29 ('88, Hasbrouck),
it remains throughout most of September" (Richmond); at
Hagerstown to September 19 ('79, Small).   On June 7 ('85) a
nest held 4 eggs slightly incubated ; on June 25 ('93) 4 birds
were nearly ready to fly ; on July 2 ('93, Wholey) a nest was
ready for eggs; on July 4 ('91) adults were feeding young
barely out of the nest, and on August 15 ('91, Gray), a pair
were leading full grown young.   At Washington it has been
found nesting along Rock Creek, where on May 31, '85, a nest

MARYLAND ACADEMY OF SCIENCES.

with 6 well incubated eggs was found, and on June 14, '85, and June 5, '87, nests with 5 young about half grown were also found (C. W. Richmond, Auk, v, 23–4).

Helminthophila pinus (641). Blue-winged Warbler.

This species, fairly numerous in migrations, from May 5 ('95) to May 16 ('93, Wholey), and from August 18 ('91, Gray) to September 21 ('94), has been found as a regular, but rare summer resident along Gwynn's Falls, between Calverton and Franklin (Blogg and Gray), where in '92, on June 11 and 14, young birds just out of the nest were noted, and on the latter date 3 were caught. On July 5, another brood had just left a nest. In '93, on May 20, a nest was half built; on the 27th it contained one egg, but later it was deserted. On June 13 a nest was found containing 3 birds just hatched and an infertile egg ; on July 1, another had 4 birds under a week old and an infertile egg, and on July 12, another nest was found with young birds. On May 29, '94, a nest was found with four fresh eggs.

On June 15, '95 (Fisher) a pair was seen at Mt. Washington where five days later I noted a pair feeding young, and on July 10, '91 (Fisher), one bird secured from a flock of 5 or 6 at Ruxton indicates its nesting there. "Geo. Marshall and others have found it breeding at Laurel, Md., and at Washington, where it is rare as a breeding bird, a nest was found by Mr. H. H. Birney early in June, '80, with 4 eggs about to hatch" (Richmond).

Helminthophila leucobronchialis (—). Brewster's Warbler.

The identity of this bird is not very clear, but it is supposed to be a hybrid ; it has been found from Virginia to Connecticut, and as far west as Michigan. On May 15, '85, one was taken near Fort Myer, Alexandria County, Va., by Wm. Palmer (Auk, ii, 304), and a very typical male was taken at Beltsville on May 1, '95, by A. H. Thayer (C. W. Richmond, Auk, xii, 307).

Helminthophila chrysoptera (642). Golden-winged Warbler.

Uncommon migrant; on May 7, '92 (Pleasants), one was taken and several others seen near Towson. On May 29, '92 (Wholey), one at Waverly, and on September 5, '93 (Gray), another near Franklintown. At Laurel a female was taken on May 8, '94, by Geo. Marshall. At Washington "one was seen on May 2, '90, and specimens were taken on May 4, '88 (Hasbrouck), and '90 (Figgins); on August 8, '89; on August 11, '89 (Palmer); and a pair on August 17, '89. There are numerous other records" (Richmond). On Dan's Mountain I noted one on June 10, '95.

Helminthophila ruficapilla (645). Nashville Warbler.

Regular, though not common migrant. On May 6, '93, and 7, '92 (Pleasants), single birds were taken near Towson. On May 11, '92 (Wholey), 3 birds were shot out of about 20 at Waverly, where on the 12th they were still more numerous; at the same place they were very numerous on May 14, '93, but 3 days later only one was noted. One was taken at Towson on September 18, '93 (Pleasants). At Washington "specimens were taken on May 1, '89 (R. Ridgway); May 3, '88 (A. K. Fisher); May 13, '82 (H. M. Smith); September 5, '82 (H. M. Smith); September 16, '94 (Figgins); and September 18, '90, etc." (Richmond).

Helminthophila celata (646). Orange-crowned Warbler.

"Migrates sparingly east of the Alleghanies" (Manual, 488). On October 12, '89, one was taken at Munson Hill, Va., a few miles from Washington, by Dr. A. K. Fisher (Auk, vii, 96), and Captain Crumb shot one at Cobb's Island in the fall of '87 (letter to W. H. Fisher). "Wm. Palmer shot one on October 14, '94, at Kensington, Md. This is the latest warbler to arrive, all eastern North American dates are very late, compared with those of the other warblers. In spring it should be looked for in March" (Richmond).

Helminthophila peregrina (647).   Tennessee Warbler.

Irregular fall migrant, has not yet been noted in spring. Specimens have been taken near Baltimore on September 18 ('93, Gray) and September 20 ('93, Pleasants). "At Washington, on August 31 ('90, Figgins; W. L. Richmond); September 28 ('82, Palmer; '90); September 29 ('89, Figgins); October 3 ('94, Palmer); October 11 ('61, T. C. Smith); October 12 ('90); November .30 ('89), and others are recorded. Some seasons it is not rare in the tall, rank growth of weeds on the Potomac Flats" (Richmond).

Compsothlypis americana (648).   Parula Warbler.

A common migrant, this species also spends the summer with us in numbers, where it finds the hanging moss (*Usnea*), and occasionally elsewhere. First noted near Baltimore on April 25 ('91, Gray), it was common on May 4 ('93, Fisher) and began to thin out by May 17 ('92, Wholey). The first fall movement was noted on August 21 ('94), and it was numerous from September 12 ('90, Gray) to October 8 ('92, Gray), the latest date being October 17 ('88, Resler). At Washington it "was first noted on April 19 ('91), and common on April 28 ('89); the last was recorded on October 16 ('87)" (Richmond).

During summer, noted near Baltimore on June 3 ('91, Resler), June 8 and 30 ('93, Wholey), July 4 and 14 ('93 Gray) and July 29 and August 5 ('92, Resler). At Washington on June 10 ('86, Richmond).

On the Isle of Wight, Worcester County, a few miles north of Ocean City, Md., on June 8, '94, I saw a pair feeding four flying young; this prompted further investigation, and on the 12th I found them very numerous, in the 4 hours I spent there, noting at least 50 pairs of adults with young, each brood keeping separate.

On Dan's Mountain I noted a female on June 5, '95, the only one seen there.

Dendroica tigrina (650). Cape May Warbler.

I have only records of 2 specimens taken near Baltimore, one on October 21 ('93, Wholey) at Waverly, the other on October 22 ('90, Resler) at Back River. "Not by any means rare in the District of Columbia, some seasons quite common; quite a number of specimens are on record between May 8 ('89, R. Ridgway) and May 16 ('89, R. Ridgway). On August 4, '92 (E. J. Brown), one was taken and they are again numerously recorded from August 25 ('90,Figgins) to October 7 ('82, H. M. Smith), while on December 16, '88 (Figgins), an adult male was shot ; it was in company with *D. coronata* and was in good health and spirits when found" (Richmond).

Dendroica æstiva (652). Yellow Warbler.

Scattered everywhere in migrations, and locally common during summer. On April 6 ('93, R. C. Watters) quite a number were seen and one shot in Dorchester County. April 15 ('95, Wm. Palmer), at Washington; April 21 ('83 Small), at Hagerstown, and April 24 ('89, Resler), are first dates. They soon become numerous, but thin out by May 11 ('92). In the fall I have never seen them so numerous as they are in spring, generally only a single bird being seen at a time, the last on October 2 ('94). Nests with eggs are recorded from May 15 ('95, Gray) to June 16 ('95). Sets are 3 of 3, 1 of 4, and 1 of 5. Common at Vale Summit, June 5–14, '95.

Dendroica cærulescens (654). Black-throated Blue Warbler.

An abundant migrant, noted as numerous from April 30 ('93) to May 15 ('92) ; the last were seen on May 28 ('94), when a pair, male and female, spent the day at Waverly, Baltimore City. August 27 ('89, Resler) notes the first return in the fall and they were numerous from August 31 ('93 Gray) to October 18 ('93), when an enormous number were in a small patch of woods. At Washington, "from April 27 ('88, Hasbrouck) to May 30 '3 3 , and from August 31 ('90) to October 19 ('92)" (Richmond).

Dendroica coronata (655).   Myrtle Warbler.

Abundant migrant, some occasionally wintering with us. On September 7, '94, a single bird was seen in Druid Hill Park. On September 30 ('94) the migration commenced and they are numerously noted to November 12 ('92, Gray) ; after which they were recorded on November 24 ('92), December 28 ('91, Gray) and January 28 ('93, Gray), becoming common again on February 4 ('93, Gray) and remaining so until May 5 ('95), the latest date being May 19 ('93, Resler). At Washington, it is "a common winter resident from September 30 ('90; '94) to May 20 ('88)" (Richmond). At Hagerstown, "plentiful in November and February, many spent the winter of '79-80, while other winters only a few were seen. In '83 they remained in considerable numbers until May 6" (Small).

During the latter part of June, '79, a male, female and 3 young about half grown were observed near Havre-de-Grace. The male was perfectly healthy, but the female had a healed up broken wing and was only able to flutter a short distance, which accounts for their staying to breed (Ludwig Kumlein, B. N. O. C., v. 182.)

Dendroica maculosa (657).   Magnolia Warbler.

Abundant migrant; noted from May 4 ('93, Fisher) to June 3 ('94, Gray), and from August 22 ('94), to October 5 ('87, Resler). At Washington "from April 22 ('91) to May 30 ('91), and from August 16 ('89) to October 5 ('90)" (Richmond). "The nest has been found in Somerset County, Pa., where I am informed they breed regularly" (Birds Pa., 283).

Dendroica cærulea (658).   Cerulean Warbler.

Rare east of the Alleghanies. On July 14, '93, an adult male and 2 young birds were taken; the male had been heard singing for a week previously (J. H.Pleasants, Jr., Auk, x, 372). On May 5, '88, a male was taken at Rock Creek, and on May 11, '90, a female on the Virginia side of the Potomac near Washington (E. M. Hasbrouck, Auk, v, 323 and vii, 291).

At Hagerstown it was noted as a rare transient and questioned as a summer resident (Small), while Dr. H. D. Mearns gives it as "a rare summer visitant" in Somerset County Pa., (Birds Pa., 269 and 284).

Dendroica pensylvanica (659). Chestnut-sided Warbler.

A common migrant from April 29 ('93, Gray) to May 25 ('92, Resler), and from August 15 ('90, Pleasants) to September 25 ('93, Gray). On October 18, '93, a remarkably large number of this species were observed. They, with other species, fairly swarmed in an 8-acre piece of woods; my note book says: "The ground was covered with Towhees and Snowbirds, the bushes were full of White-throats, Song Sparrows and Snowbirds, while the trees were filled with Black-throated Blue, Black-throated Green, and Chestnut-sided Warblers, Golden and Ruby-crowned Ringlets, Robins and Blue Jays,—evidently a first-class bird-wave."

At Washington they are recorded from "April 28 to May 30 ('91), and from August 10 ('89; '94, Figgins) to September 29 ('89)" (Richmond).

On July 4, '93 (Wholey), a pair were taken near the Blue Mountain House, and on July 24, '93 (Gray), a pair with "worms" in their bills were seen within 100 yards of High Rock, showing that they breed on the Blue Ridge. On Dan's Mountain they were fairly common from June 5 to 14, '95; on the 9th I collected a nest with 4 eggs about half incubated.

Dendroica castanea (660). Bay-breasted Warbler.

Migrant, not common ; specimens have been taken as follows: May 16 ('91, Blogg) at Franklintown, May 21 ('93, Wholey) at Waverly, May 24 ('90, Pleasants) at Towson, and May 28 ('93, Wholey) at Waverly; September 20 ('93, Gray) at Calverton, September 21 ('93, Pleasants) at Towson, and September 28 ('89, Resler) at Back River.

At Washington "a late arrival in spring of irregular abundance, usually uncommon ; from May 10 to 27 ('88; 5 shot by Wm. Palmer), and from August 29 ('87, R. Ridgway) to October 19 ('88, R. Ridgway). On September 22, 24 and 29, '89, they were common at Great Falls, Md." (Richmond).

Dendroica striata (661).   Black-poll Warbler.

Common migrant from May 4 ('92, Resler) to May 28 ('87, Resler; '93, Wholey), and from September 24 ('84, Resler) to October 18 ('90, Resler).   At Sandy Springs, to June 3 (Stabler).   At Washington "numerous from May 1 to June 1 ('88, '89, and '90, Figgins ; '91), and from September 1 ('89 and '94, Figgins) to October 20 ('89)   One was seen on June 3, '89; one was shot on June 4, '58, (Coues), and one was heard singing on June 12, '94 (Brown)" (Richmond).   "On July 30, '93, an adult male was shot at Washington (E. J. Brown, Auk, xi, 79).

Dendroica blackburniæ (662).   Blackburnian Warbler.

Common migrant from May 2 ('93, A. Wolle) to May 19 ('92, Resler), and from August 30 ('93, Gray) to October 13 ('83, Resler).   At Washington "as early as August 19 ('94, Brown)" (Richmond).   On July 28, '92, in Howard County, a young male was taken by A. Resler (Transactions of the Maryland Academy of Sciences, 1892, 203).   On June 10, 95, on Dan's Mountain, while watching a pair of Louisiana Water Thrushes, a male flew close to me and commenced bathing.

Dendroica dominica (663).   Yellow-throated Warbler.

"Southern Atlantic States, north regularly to Maryland (near seacost)" (Manual, 503).   On May 12, '95, at Fairview, a fishing shore on Rock Creek, Anne Arrundel County, and 9 miles from Baltimore, Mr. W. H. Fisher noted 3 birds of this species; two were together in the woods, the third was in a tree at the boat-landing.

"At Washington it is not common ; specimens have been taken as follows: July 15, '94, 3 (Brown and Palmer); July 20, '90, 3 or 4 (Figgins and Richmond); July 27, '90, 1; July 28, '89, 4 (Figgins and Richmond); July 29, '94, 1 (Brown and Palmer); August 1, '89, 5; August 1, '94, 4 (Figgins); August 5, '89, 1; September 4, '90, 1 seen. In '93, at Johnson's Gully, near Marshall Hall, Md., where it undoubtedly breeds, Palmer and Hasbrouck found it on June 1 and 11, July 22 and August 10. On Smith's Island, Va., about May 24, '94, I saw one with a ' worm ' in its mouth, evidently feeding young " (Richmond).

Dendroica virens (667).   Black-throated Green Warbler.

Common migrant from April 21 ('93, Blogg) to May 15 ('92), and from August 30 ('90, Resler) to October 22 ('90, Pleasants). On October 18, '93, they were remarkably numerous in the bird wave already noted. At Washington from April 22 ('88, M. M. Green) to May 19 ('88; '95, Palmer), and from August 26 ('89, Figgins, at Jefferson, Md.) to October 21 " (Richmond). On June 14, '95, on Dan's Mountain, a pair were feeding flying young.

Dendroica kirtlandi (670).   Kirtland's Warbler.

Migrates through southeastern United States and Mississippi Valley; summer home unknown, winters in the Bahamas. " Wm. Palmer shot one on September 25, '87, near Fort Myer, Va. (about one-quarter mile from the Potomac River)" (Richmond).

Dendroica vigorsii (671).   Pine Warbler.

Fairly numerous during migrations, this species, local in its distribution, spends the summer with us in limited numbers. Numerously noted from April 9 ('90, Resler) to April 25 ('85, Resler), and from August 3 ('93, Resler) to October 22 ('92, Blogg). On June 29, '93 (Resler), one was taken at Back River, and on July 2, '93, another in Dulaney's Valley. At

Washington "noted from March 24 ('89, Figgins) to October 25; it is a rare summer resident but swarms at times in the fall. Figgins found them in immense numbers on August 31, '90, and shot 18" (Richmond).

In Dorchester County, on April 6, '93 (R. C. Watters), several were seen and one taken. In Worcester County, on June 12, '94, on the mainland, about a mile from Ocean City, I watched a male for about an hour in hopes of locating the nest which evidently was close at hand, but I did not find it.

Dendroica palmarum (672).   Palm Warbler.

"Occasional (or casual) during migrations east of the Alleghanies" (Manual, 517). The following specimens have been taken near Washington, where it is "probably a regular, though rare, migrant." April 22,' 85, Roslyn, Alexandria County, Va.; April 29, '88, Roslyn, (C. W. Richmond); May 6 and May 11, '89, Laurel, Prince George County, Md. (R. Ridgway); May 11, '90, Riverdale, Prince George County, Md. (C. W. Richmond); May 11, '81, Soldiers' Home, District of Columbia (L. M. McCormick); September 18, '87, Potomac Landing, Alexandria County, Va.; September 22, '93, Four Mill Run, Alexandria County, Va. (J. E. Brown); October 4, '91, Ballston, Alexandria County, Va. (Wm. Palmer, Auk, xi, 333).

Dendroica palmarum hypochrysea (672a),   Yellow Palm Warbler.

A regular migrant; common, usually appearing in small flocks, though often single birds are seen, from April 4 ('93, Gray) to April 29 ('91, Resler), and from September 12 ('95, Hoen) to October 22 ('93). At Washington "from March 31 ('89) to April 29, and in the fall they were still common on October 19 ('90, Figgins)" (Richmond). "One year they were very common at Hagerstown in October, swarming everywhere, some even coming in at the windows" (Small).

Dendroica discolor (673). Prairie Warbler.

A summer resident, but so decidedy local that until its special haunts are discovered its presence is not suspected. These are usually of limited extent, but quite a number are known near Baltimore, a few pairs breeding at each. Noted from April 22 ('91, Resler) to September 5 ('83, Resler). On June 4 ('90, J. H. Fisher, Jr.) a nest held slightly incubated eggs, and on July 16 (93) young birds left a nest on being disturbed. Sets are 3 of 3, and 1 of 4. "Common at Washington from April 19 ('91) to September. Mr. Figgins found a nest with 4 eggs on May 14, '91, and another with 3 on May 30, '88" (Richmond).

Seiurus aurocapillus (674). Oven-bird.

Common summer resident; first noted on April 9 ('93), and numerous from April 21 ('93, Resler) to September 16 ('94); the last was recorded on October 18 ('90, Resler). Extreme nesting dates are May 24 ('91, J. H. Fisher, Jr.), fresh eggs, and August 6 ('93), young just out of the nest. Sets are 2 of 3, 7 of 4, and 1 of 5. At Washington "to October 17, '90; on May 20, '88, a nest with 4 eggs was found" (Richmond).

Seiurus noveboracensis (675). Water-Thrush.

Common migrant from April 27 ('92, Resler) to May 25 ('93, Resler), and from August 29 ('94, Resler) to October 16 ('92, Wholey). At Washington "from April 22 ('94, Wm. Palmer) to May 25. On July 21 ('94, Figgins) one was found that had killed itself by flying against telegraph wires, and it is numerously recorded from that on until September" (Richmond).

Seiurus noveboracensis notabilis (675a). Grinnell's Water-Thrush.

A western species. Two specimens taken in Virginia, near Washington, May 11, '79, and May 5, '85, are in the collection of Mr. W. Palmer (Auk, v, 148). Another was captured by Dr. A. K. Fisher, August 5, '86, not far from the Long Bridge, on the Virginia side of the Potomac" (Birds Vas., '85–6.)

Seiurus motacilla (676).   Louisiana Water-Thrush.

Common during migrations. This species probably spends the summer with us in greater numbers than is generally supposed. Noted from April 3 ('93, Gray) to September 24 ('92, Blogg). On June 13 and 22, '93 (Gray), a pair were seen feeding young near Franklintown. At Washington "a nest and 5 badly incubated eggs were obtained on Piney Branch, May 25, '88, by G. E. Mitchell. Wm. Palmer got a fully fledged young bird about June 14, '91 " (Richmond). On June 10, '95, I found two pairs on Dan's Mountain, where, no doubt, they were nesting.

Geothlypis formosa (677).   Kentucky Warbler.

Summer resident, not rare round Baltimore; a pair usually occupying each marshy spring head in heavy woods. Noted from April 29 ('92, Blogg) to September 8 ('95); nesting dates range from June 18 ('93), young just hatched, to July 27 ('95), young still in nest. Sets are 1 of 3 and 4 of 4. At Washington "not very common, May 3 to September 5. On June 15, '79, Mr. H. W. Henshaw found a nest with 4 eggs somewhat incubated" (Richmond). At Johnson's Gully, Md., on May 30, '92, a nest with 5 eggs was found, and on June 6, '29, another with 4 eggs, both slightly incubated (E. M. Hasbrouck, Auk, x, 92).

Geothlypis agilis (678).   Connecticut Warbler.

A rare migrant. I have no spring dates, but specimens have been taken from September 20 ('93, Pleasants) to October 1 ('90, Resler). At Washington "very rare in spring. Mr. L. McCormick shot one at Falls Church, Va., in May, '79, and Wm. Palmer got a female on May 23, '91. It is fairly common in fall, and has been taken from August 28 ('86, Dr. A. K. Fisher) to October 12 ('90), when two were taken on the Potomac Flats and another seen " (Richmond).

Geothlypis philadelphia (679). Mourning Warbler.

Rare migrant; near Baltimore single birds have been recorded as follows : May 28, '91, Towson (Pleasants), shot ; May 26, '77, Back River (Resler), shot; June 3, '93, Franklin Road (Gray), seen ; August 24, '87, Patapsco Marsh (Resler), shot; August 30, '93, Back River (Resler), shot ; September 1, '90, Franklin Road (Blogg), seen ; September 13, '79, Bay View (Resler), shot ; September 18, 93, Franklin Road (Gray), seen ; September 26, '94, Back River (Resler), shot; September 30, '93, Dulaney's Valley (Wholey), shot ; October 5, '87, Patapsco Marsh (Resler), shot.

On August 17, '94, an immature female was taken at Laurel, by Mr. Geo. Marshall. At Washington "very rare from the middle to the end of May and from August to October 1, '94, when Wm. Palmer got an immature female" (Richmond).

Geothlypis trichas (681). Maryland Yellow-throat.

Our most abundant summer warbler; first noted on April 20 ('91, Resler), it soon becomes common and remains so until October 6 ('95), the last being recorded on October 19 ('92, Resler). The nests of this species, as of all ground buildings birds, are difficult to find, so I have but few items ; extremes are May 24 ('92, Wholey), a nest ready for eggs, and August 13 ('93), when young leave a nest. Sets are 1 of 3, and 2 of 4. At Washington "abundant from April 15 ('91, R. Ridgway) to October 21 ('95, Palmer and Matthews)" (Richmond). Not very common on Dan's Mountain, June 5–14, '95.

Icteria virens (683). Yellow-breasted Chat.

Common summer resident, from April 30 ('92, Gray); they leave early and but few are seen after August 6 ('93), though single birds have been noted until August 30 ('93, Gray). At Washington "from April 29 ('74, Palmer, Hasbrouck and Preble ; '88, Hasbrouck) to September " (Richmond). It is given to September 20 in Pennsylvania (Birds Pa., 299). Eggs

·are noted from May 22 ('89, J. H. Fisher, Jr.) to July 12 ('91).
Sets are 9 of 3, 2 of 4, and 1 of 5. A few pairs were on Dan's
Mountain; a nest contained 3 eggs on June 14, '95.

Sylvania mitrata (684). Hooded Warbler.

A rare migrant; near Baltimore specimens have been taken :
May 3, '90, and 10, '84 (Resler); May 15, '92 (Wholey) ;
September 8, '93 (Gray); September 20, '90 (Resler), and
one was seen on September 22, '95. Possibly it breeds in limited
numbers in Howard County, for on August 4, '94, near Botterill P. O., 14 miles southwest of Baltimore, Mr. A. Resler
collected a young bird not long out of the nest, and four days
later a pair of adults, male and female. At Washington "a
rare migrant, may breed. Specimens have been taken from
May 1 ('88, Hasbrouck) to the end of the month; and from
August 19 ('94, Figgins), at Kinsington, to September 15 ('90)"
(Richmond).

Sylvania pusilla (685). Wilson's Warbler.

A fairly common migrant from May 6 ('93, Gray) to May
26 ('95), and from September 11 ('92, Gray) to September 23
('93, Wholey). At Washington "rather common from May 8
to May 23 ('91), and from August 31 ('90) to the middle of
September" (Richmond).

Sylvania canadensis (686). Canadian Warbler.

Common migrant from May 4 ('95, Fisher) to May 29 ('94,
Resler), and from August 7 ('89, Resler) to September 18 ('90,
Gray). At Washington "from May 5 ('94, Hasbrouck) to
May 27, and from August 7 ('87) to September 24 ('89)"
(Richmond). On Dan's Mountain, on June 10, '95, I came
across a male singing, and later a pair with young in the nest.

Setophaga ruticilla (687). American Redstart.

Common migrant from April 25 ('91, Blogg) to May 24
('93, Wholey), and from August 24 ('94) to October 15 ('93);
most of the notes, after the middle of September, record imma-

ture birds. It is also fairly common, though local, during the summer. A pair were seen mating on May 15 ('92), and young birds out of the nest are noted from June 3 to July 18 ('93, Gray). On June 20, '93, Mr. W. H. Fisher and I found them very numerous at Mount Washington. In a walk of about three-fourths of a mile we noted 9 pairs feeding young, and no doubt there were others we did not see. Sets are 3 of 3. At Washington it is "a common migrant, first noted on April 19, '91. A few breed. Hasbrouck found a nest with 4 eggs nearly hatched on May 21, '88" (Richmond). On Dan's Mountain, June 5 to 14, '95, they were very numerous; their numbers about equalled that of all other birds seen there. On the 9th young birds left a nest before I could count them.

Family MOTACILLIDÆ—Pipits.

Anthus pensilvanicus (697). American Pipit.

Common in flocks from October 13 ('95) to November 19 ('94), and from February 11 ('94) to May 13 ('92); occasionally some winter with us. At Washington "from October 15 ('90) to May 4 ('89); more numerous in October, November and March and April than in midwinter. They were common until November 30, 90" (Richmond).

On November 12, '93, I found a large flock in Dulaney's Valley; some were running and presumably feeding on pasture, others on a part that had been newly turned up, while the greater number were either bathing in a shallow puddle or dressing their feathers in an adjoining tree, being perched on the branches from the ground up to the top, fully 50 feet from the ground; they, however, were more at home on the fence rails or on the ground.

Family TROGLODYTIDÆ—Thrashers, Wrens, etc.

Mimus polyglottos (703). Mockingbird.

Resident in the southern counties of Maryland and regular in summer as far north as Kent and Anne Arundel Counties. In the balance of the state it can only be called a straggler. On May 2, '93 (Fisher), one was at Glyndon. On June 7 and 14, '85, one was singing in great style at Fork. On September 17, '93, two were fussing in a tree in Dulaney's Valley, where Mr. Thomas Peerce informs me a pair "used" until a few years ago, and also that another pair regularly spent the summer just across the ridge in Long Green Valley, until about the same year. In the spring of '82 (Pleasants) two nests with eggs were found near Towson.

On June 28, '93 (Blogg), a nest with young about a week old was found at Bay Ridge, Anne Arundel County. In Queen Anne County, on May 30, '92, a nest was ready for eggs, and on May 27 and 31, and June 1 and 2, nests were found with 3 young birds in each. At St. Michael's, Talbot County, 4 fresh eggs were found on June 13, '94.

Galeoscoptes carolinensis (704). Catbird.

Common summer resident from April 25 ('92, Gray) to October 22 ('90, Resler). One was noted in Dulaney's Valley on December 24 ('94, W. L'Allemand), and on April 2 ('93) two very noisy birds spent the day at Waverly. At Washington "from April 16 ('94, R. Ridgway) to October ; Wm. Palmer has taken a specimen in December, and on January 13, '89, I found the remains of one that had been killed by some bird of prey" (Richmond). On May 4 ('92) a pair were building, and eggs are recorded from May 16 ('91) to August 16 ('91). Sets are 1 of 1, 6 of 2, 31 of 3, 15 of 4, and 2 of 5. At Vale Summit, I only saw 3 in 10 days.

The following is the result of observations made on a nest built in a veranda at Tacoma, D. C. After the nest had been completed 24 hours, the first egg was laid on May 11, '93, at 10.35 A. M., the second at 9.40 A. M., the third at 9.15 A. M., and the fourth at 10.15 A. M. on successive days. For the first 4 days she sat upon them only at irregular intervals, but this habit soon changed after that time. On May 25 there were no birds hatched at dark, but on the morning of the

26th 3 young were in the nest, the other hatched the next night. At 6.45 P. M. on June 5 they all left the nest together (R. W. Shufeldt, Auk, x, 303–4).

Harporhynchus rufus (705).   Brown Thrasher.

Common summer resident, from April 19 ('94) to September 21 ('94); extreme dates are April 7 ('90, Resler) and November 5 ('93). "At Laurel, on April 2 ('88, R. Ridgway), two were seen.   At Washington from April 6 ('88) to November 13 ('87)" (Richmond).   May 21 ('93), a nest with birds about one-half grown, and August 6 ('93), another with birds only 3 or 4 days old, are extreme nesting dates. Sets are 2 of 2, 4 of 8, 5 of 4, and 1 of 5.   Usually nesting in a brier tangle or thick bush, on May 29, '81, one was placed on the broken-off top of an appletree stub about 7 feet high.   On July 26, '91, one was on the 4th rail of a worm fence, and on June 24, '91, another was on the ground under a potato vine.   At Vale Summit they were not common; on June 7 a nest held 4 fresh eggs.

Thryothorus ludovicianus (718).   Carolina Wren.

Common resident, occasionally vying with the House Wren in semi-domesticity.   On April 8 ('94) a nest was nearly finished, and on May 2 ('90) a brood of young left a nest, while as late as August 10 ('93) slightly incubated eggs were taken. Sets are 1 of 3, 1 of 4, 2 of 5, and 2 of 6.

On April 26, '91, I was shown a nest containing 3 fresh eggs placed behind an ornament in the corner of a friend's parlor; later, I was told that either five or six young were raised there. At Vale Summit one bird was seen on June 13, '95.

Thryothorus bewickii (719).   Bewick's Wren.

A rare bird east of the Alleghanies.   Single birds are recorded at Washington as follows: April 5, '92 (R. Ridgway), seen; April 6, '83 (Wm. Palmer), shot; April 8, '88 (M. M. Green), shot; April 10, '82 (Wm. Palmer), shot; April 22, '88 (M. M. Green), seen; November 24, '89 (J. D. Figgins), shot; December 22, '90 (C. W. Richmond), seen (Robt. Ridgway, Auk, ix, 307, and Richmond). Possibly it is common in summer in parts of western Maryland. On July 4, '93 (Wholey), about one hundred yards from Quir-Auk, on the Blue Ridge Mountain, a nest of this species was found in a hollow log ; in it were 4 young, which fluttered out, leaving 2 infertile eggs. On the same day a specimen was secured near the Blue Mountain House. On August 2, '93 (Gray), one was observed near Hagerstown. On June 9 and 14, '95, I noted this species at Vale Summit, and on the 11th a pair were seen with a brood of young.

Troglodytes aëdon (721).   House Wren.

A common summer resident. This semi-domesticated, wholly independent and irrepressibly impudent little bird has been noted from April 12 ('95, Fisher) to October 10 ('93), on which latter date a number were seen at Bright Lights, near North Point. Possibly they may arrive earlier than above given, for on April 27 ('93) a nest was about one-half built, and 3 days later another was nearly ready for eggs. The earliest date, however, I have found eggs is May 19 ('95), the latest July 17 ('92), and on August 26 ('94) young birds barely out of the nest were seen. Sets are 2 of 3, 3 of 4, 7 of 5, 4 of 6, and 2 of 7.

This species often pre-empts the nests of other birds, the Downy Woodpecker being the most usual victim. On May 26, '94, a Carolina Wren's nest, which I had watched being built, held one egg; on June 3 a House Wren was found it it, and having added 5 of her own was sitting on them, all 6 being slightly incubated. As a rule, when the House Wren jumps a claim it modifies things to suit itself, but in this case it made no changes, the nest being exactly as the Carolina Wren had built it.

Troglodytes hiemalis (722).   Winter Wren.

Common winter resident from September 26 ('94, Resler) to May 7 ('93, Blogg, Fisher and F. C. K.). At Washington

"from September 27 ('89, R. Ridgway) to April 29 ('88)"
(Richmond). While a few come round the farm buildings,
others may be found in dry upland woods, but to see any number in the course of a day's walk, our wooded watercourses
must be followed, where they will usually be found among the
roots that hang from the washed banks, the individuals being
about one-half a mile apart. On November 27, '92, Mr.
W. N. Wholey and I heard one sing in Dulaney's Valley.

Cistothorus stellaris (724).   Short-billed Marsh Wren.

Very rare. On April 18, '79, a specimen was presented to
the Maryland Academy of Sciences by Mr. W. S. Clayton.
Presumably it was taken in the vicinity of Baltimore. At
Washington E. M. Hasbrouck has taken two specimens; one, a
female, on May 9, '90 (Auk, vii, 289), the other on May 3,
'93 (Richmond). " Rare migrant, and in southern New Jersey
occasional (regular?) winter resident" (Birds E. Fa. and N. J.,
144; Auk, ix, 204).

Cistothorus palustris (725).   Long-billed Marsh Wren.

Common summer resident in the marshes of tidewater Maryland ; it probably also winters in limited numbers in southern
Maryland, as Messrs. S. N. Rhoads and Witmer Stone found it
" tolerably common in cattail swamps" at Cape May City, N.
J., January 26 to 29, '92, (Auk, ix. 204). Noted near Baltimore from April 28 ('94, Wholey) to October 14 ('91, Resler),
and at Washington to October 19 ('90, Richmond). Nests with
eggs range from June 3 ('90) to August 7 ('90), on which latter
date 3 other nests were ready for eggs, so presumably they nest
later. Sets are 1 of 2, 1 of 3, 9 of 4, 6 of 5, and 1 of 6. At
Hagerstown one was observed " for a few days in spring in
the yard" (Small).

Family CERTHIIDÆ—Creepers.

Certhia familiaris americana (726).   Brown Creeper.

Fairly common during winter from September 26 ('94, Resler) to May 8 ('75, Resler). " One was taken at Washington

on September 22, '55, by Dr. Coues" (Richmond). On September 27 ('94, Fisher) one was seen in a tree at the corner of Park Avenue and Madison Street. At Hagerstown "from October to May" (Small).

Family PARIDÆ—Nuthatches and Tits.

Sitta carolinensis (727).   White-breasted Nuthatch.

Common resident, but usually seen singly or in pairs. On March 31 ('94, Blogg) a pair were found building, and on April 7 the nest contained 1 egg. On July 26 ('94) young just out of the nest were observed.

On June 11, '95, a single bird was noted on Dan's Mountain, and on July 6 ('95, Tylor), a pair were feeding young at Deer Park.

Sitta canadensis (728).   Red-breasted Nuthatch.

Common during some winters, this species is either wanting, or very rare during others.   During the severe winter of '92–3, they were quite numerous round Baltimore.   Usually found singly, at times several may be seen fairly close together.   September 11 ('80, Resler) and May 6 ('93, Gray), are extreme dates; on this last day they were quite numerous and a number of specimens were taken.   At Washington to May 10 (Richmond), and at Hagerstown to May 4 (Small).

Sitta pusilla (729).   Brown-headed Nuthatch.

Apparently a regular summer resident in southern Maryland. On May 28, 92, in Queen Anne County, I found a pair building a nest in a small hole in a dead pine stub, and later in the day, about three-quarters of a mile distant, I came across another pair. In Worcester County on the mainland, about a mile from Ocean City, on June 8, '94, I found a single bird.

"It is common at St. George's Island and Piney Point, Md., near the mouth of the Potomac, and on Smith's Island, Va. Between Cape Charles and Brighton it was noted in small numbers and is probably common" (Richmond).

Parus bicolor (731). Tufted Titmouse.

Common resident. On April 10 ('92) a nest was ready for eggs, and on August 4 ('94) young not long out of the nest, keeping in a close bunch and still being fed, were seen. Sets are 1 of 5 and 1 of 8. "Their notes are heard loudest on clear winter days, and at Hagerstown they are locally called 'Storm Bird'" (Small). On Dan's Mountain I noted single birds on June 11 and 12, '95.

Parus atricapillus (735). Chickadee.

Irregular winter visitant, sometimes common in cold seasons ('92–3 for instance), when the majority of the following species leave us. Noted from October 15 ('92; '93) to March 21 ('95).

"Mr. Henry Marshall has taken it at Laurel" (A. C., 37). "This bird was very abundant about Washington during March and April, '85. . . . Owing probably to the severe winter they were driven south, returning about the middle of March; the first specimens were taken March 15, and others were taken every week until April 19, when 6 were shot and many others seen. The weather during April was fine and warm, and the birds were singing and apparently quite at home. But few *P. carolinensis* were seen until the last week of April, showing that they too had been driven much further south" (Wm. Palmer, Auk, ii, 304). Noted at Hagerstown as "common during the winter of '80–81"; none, however, were seen the previous or the two succeding seasons (Small).

Parus carolinensis (736). Carolina Chickadee.

Common resident. On April 23 ('93) a nest held 2 fresh eggs, and on July 23 ('93) young, not long out of the nest, were seen. Sets are 2 of 2, 1 of 3, 2 of 4, 1 of 5, 1 of 6, and 2 of 7.

At Hagerstown "resident, but scarce in summer" (Small). On Dan's Mountain, June 6, '95, young were in the nest of the only pair seen.

Family SYLVIIDÆ—Kinglets and Gnatcatchers.

Regulus satrapa (748).   Golden-crowned Kinglet.

Winter resident; very common during some seasons ('92–3, for instance) and comparatively rare others. They are, however, most numerous during migrations, thinning out about the end of October and getting numerous early in March. Extreme dates are September 30 ('93, Gray) and April 22 ('93, Gray). At Washington "one was taken on April 27 ('88, Hasbrouck)" (Richmond), and at Hagerstown it was noted as "staying until May" (Small).

Regulus calendula (749).   Ruby-crowned Kinglet.

A common migrant; possibly a few winter with us during mild seasons. Noted from September 26 ('94, Resler) to November 5 ('92, Gray), and from April 2 ('87, Resler) to May 7 ('93). At Hagerstown, a "common migrant in March and April, October and November; plentiful in fall, but scarce in spring" (Small). At Washington "from September 25 ('87) to early in November, and from April 8 ('88) to May 10 ('91). Specimens have been taken in winter: two on December 1 ('89, C. W. Richmond; A. K. Fisher), one on December 15, '89, and 2 others seen, and 1 on February 9 (91, Figgins)" (Richmond). "On December 5, 92, I saw and positively identified a single Ruby-crowned Kinglet in the grounds of the Department of Agriculture, and also saw what was presumably the same individual on January 5, 6 and 14, '93, this period covering some of the most severe weather ever known here" (E. W. Clyde Todd, Auk, x, 206).

Polioptila cærulea (751).   Blue-gray Gnatcatcher.

Fairly common during summer in restricted localities, from April 1 ('93, Gray) to September 14 ('93, Gray). At Washington "from April 5 to September; one was shot on November 23, '90" (Richmond). On May 7 ('93), a nest was nearly finished; May 14 ('93, Blogg) and June 4 ('95, Fisher) are extremes for eggs. Sets are 7 of 4.

On June 12, '94, on the Isle of Wight, near Ocean City, I saw two pairs with young just out of the nests.

Family TURDIDÆ—Thrushes, Bluebirds, etc.

Turdus mustelinus (755).   Wood Thrush.

Common resident from April 29 ('93 ; '94) to October 2 ('95). Extreme dates are April 13 ('92) and October 15 ('93). At Washington "several were seen on April 19, '91, and they were common the next day" (Richmond). At Vale Summit they were fairly numerous, June 5 to 14, '95.

Eggs are recorded from May 18 ('91; '95) to July 28 ('95). Sets are 1 of 1, 18 of 2, 30 of 3, and 16 of 4.

Turdus fuscescens (756).   Wilson's Thrush.

Fairly common migrant, from April 23 ('93, Fisher) to May 30 ('92, Blogg) and from August 29 ('93, Gray) to October 6 ('94). At Washington "from April 26 to May 28, and from August 18 ('89, Figgins) to the end of September" (Richmond). At Hagerstown it was noted in October ('79, Small). On July 6, '95 (Tylor), a pair were feeding young at Deer Park.

Turdus aliciæ (757).   Gray-cheeked Thrush.

Regular, but not very common migrant, from May 12 ('92, Wholey) to May 30 ('92, Resler), and from September 12 ('94, Resler) to October 16 ('92, Wholey). At Washington "rather common from May 10 to June 5, and from September 10 to October 10" (Richmond).

Turdus aliciæ bicknelli (757a).   Bicknell's Thrush.

Apparently a rare migrant. "At Laurel Mr. Robert Ridgway shot an adult male on May 14, '88, and 4 days later an adult female ; near Washington on October 3, '85, I secured an adult male" (Richmond).

Turdus ustulatus swainsonii (758a).   Olive-backed Thrush.

Common migrant from April 30 ('92, Wholey) to May 28 ('93, Wholey), and from September 16 (92, Pleasants) to October 13 ('94). At Washington "until Oct. 20, '90" (Richmond).

Turdus aonalaschkæ pallasii (759b).   Hermit Thrush.

A common migrant from October 4 ('94) to November 28 ('91, Resler), and from March 23 ('89, Resler; 94, Fisher) to May 21 ('92, Gray). It also stays with us sparingly during winter, single birds being taken as follows : December 24, '92 (Wholey); January 1, '92 (Wholey); January 8, '87 (Resler); January 17, '92 (Wholey); January 20, '93 (Gray) ; January 27, '95 (F. C. K.), 1 seen; January 29, '80 (Resler), 1 seen; February 6, '76 (Resler), 1 seen ; February 22, '92 (Gray), 1 seen; and March 5, '90 (Resler). In Somerset County they were common from November 13 to 22, '94 (Fisher). On July 9, '90, an adult female was shot in Howard County, about fifteen miles southwest of Baltimore City, by A. Resler (Forest and Stream, xxxv, 11).

Merula migratoria (761).   American Robin.

Resident, abundant in flocks during the migrations, it is commonly scattered over the country during summer, and a few scattered flocks as a rule winter with us in the uplands north of Baltimore, while down the "necks" they are more numerous, and in southern Maryland large flocks are characteristic of the winter season. Eggs are noted from May 5 ('95) to July 30 ('81). Sets are 7 of 3 and 10 of 4. At Hagerstown eggs were also noted on May 5 ('82, Small). Not numerous at Vale Summit, June 5 to 14, '95.

Sialia sialis (766).   Bluebird.

Common resident. On March 18 ('94) a nest was ready for eggs, and birds were in another on August 12 ('94). Sets are 2 of 2, 3 of 3, 5 of 4, and 7 of 5.

In the fall and winter they often go in flocks of 5 or 6, to about 20, but generally they will be seen in pairs. On January 16, '95, Bluebirds were seen at Cumberland, where the rivers and creeks had been frozen over from December 25 (Howard Shriver, Maryland State Weather Service, Report of February, '95). On June 11, '95, a brood of flying young were noted at Vale Summit.

The latter part of the winter of '94-5 was remarkable for its severity all over North America, and the birds were thinned out in great numbers. Perhaps from being more familiarly known than other small birds, the almost complete absence of Bluebirds and Phœbes caused considerable comment. The blizzard of early February completely covered Maryland with 2 feet or more of snow, and travel was stopped for about 10 days by the drifts. Bluebirds had been numerous around Baltimore up to this time, and 2 were still alive on February 10. Later in the season a number of Screech Owl holes were found more or less filled with Bluebird feathers, and the owls were absent. Only a very few scattered Bluebirds were noted during summer and fall until November 3, when my note book says, "scattered everywhere over Dulaney's Valley in pairs, evidently a migration flight"; on November 9 a flock of about 20 was seen. At Washington "21 dead birds were found by Robt. Ridgway in a bird box on his grounds, where they had been frozen by the blizzard which almost exterminated them in this part of the country", (Richmond).

Larus atricilla (58), Laughing Gull.   About 100 were below Fort McHenry on April 23, '95, and about a dozen were in the lower harbor on October 24, '95.

Hydrochelidon nigra surinamensis (77).   "The Black Tern at Washington, D. C.   September 18, '93 I shot 13 Black Terns. Previous to this I am aware of only one recorded instance of its occurrence, one being found dead September 18, '82.   Edward J. Brown" (Auk, xi, 73).

Phalacrocorax dilophus (120), Double-crested Cormorant.   7 Cormorants, presumably of this species, were seen at Ocean City on September 23, '94 (Tylor).

Rallus virginianus (212), Virginia Rail.   Occasional in winter.   One was shot near Baltimore, late in December, '79 (Resler), and a male, in full plumage and in fine condition, was taken in a terrapin trap on January 20, '91, near Easton ("Sink-boat", Forest and Stream, xxxvi, 44).

Gallinago delicata (230), Wilson's Snipe.   On December 29, '94, a female was shot at Westover, Somerset Co., by Mr. M. T. Sudler.

Tringa alpina pacifica (243a), Red-backed Sandpiper.   Three were at Lock Raven on September 29, '95.

Totanus melanoleucus (254), Greater Yellow-legs.   On November 13, '75, one was shot at Back River (Resler).

Totanus flavipes (255), Yellow-legs.   Numerous at Loch Raven from September 22 to October 13, '95.

Bartramia longicauda (261), Bartramian Sandpiper.   One was shot on September 13, '95, near Relay, by Mr. C. Gamble Lowndes.

Asio accipitrinus (367), Short-eared Owl.   On November 13, '95 one was shot at Bear Creek, Baltimore Co.

Coccyzus americanus (387), Yellow-billed Cuckoo. One was seen at Pikesville on October 27, '95 (Fisher).

Carpodacus purpureus (517), Purple Finch. Should be recorded to May 21 ('92 Gray); not to May 31 ('93 Fisher).

Calcarius lapponicus (536), Lapland Longspur. "The blizzard drove half a dozen of these birds into Baltimore City, where on February 4, 5, and 10, '95, they were seen in company with English Sparrows feeding in the bed of North Avenue near Caroline St. They allowed of quite close approach" (Resler).

Ammodramus maritimus (550), Seaside Sparrow. "A specimen was collected some 9 or 10 years ago, on Miller's Island, Baltimore Co., by the late Mr. Wolle, Sr., at whose house I had the opportunity to examine the bird before it was mounted" (Resler).

ERRATA.

Page 255, line 24, for Cantter read *Cantter*.
Page 269, line 19, for nivialis read *nivalis*.
Page 270, line 19, for Hutchin's read *Hutchins's*.
Page 274, line 1, for Ardella read *Ardetta*.
Page 275, line 29, for cœrulea read *cærulea*.
Page 277, line 24, for vialaceus read *violaceus*.
Page 281, line 5, for light house read *life-saving station*.
Page 292, line 14, for Black-billed read *Black-bellied*.
Page 301, line 27, for (399) read (*339*).
Page 318, line 31, for (622a) read (*466a*).

November 15, 1895.

INDEX.

A. C.—Avifauna Columbiana, being a list of birds ascertained to inhabit the District of Columbia, etc., by Elliott Coues and D. W. Prentiss. Second edition, 1883. (This includes the first edition, 1861, and the gist of P. L. Jouy's list of 1877.)

A. O. U.—The American Ornithologists' Union, Check List of North American Birds, 1886. (This is the present standard of nomenclature.)

Andubon.—The Birds of America, by John James Audubon, Vols. I to VII, 1840-44.

Auk.—The Auk, a quarterly Journal of Ornithology. (The official organ of the American Ornithologists' Union. Vol. I, No. 1, appeared January, 1884.)

B. N. O. C.—Bulletin of the Nuttall Ornithological Club. (Vol. I, No. 1, appeared in January, 1876. In 1884 it was continued as the Auk.)

Bendire.—Life Histories of North American Birds, with special reference to their breeding habits, and eggs, by Chas. Bendire, 1892. (Gallinaceous Birds, and Birds of Prey.)

Birds E. Pn. and N. J.—The Birds of Eastern Pennsylvania and New Jersey, by Witmer Stone, 1894.

Birds N. W.—Birds of the North-West, by Elliott Coues, 1877.

Birds Pn.—Report of the Birds of Pennsylvania, by B. H. Warren. Second edition, 1890.

Birds Vas.—A Catalogue of the Birds of the Virginias, by Wm. C. Rives, October, 1890.

Chapman.—Handbook of the Birds of Eastern North America, by Frank M. Chapman, 1895.

Fisher's Hawks and Owls.—The Hawks and Owls of the United States in their Relation to Agriculture, by A. K. Fisher, 1893.

Forest and Stream.—A weekly Journal of the Rod and Gun. (Vol. XXVI, No. 1, January 27, 1886, is the first number I have access to. Two volumes each year. No. 1 bearing date of the fourth Thursday of January or July.)

Key.—Key to the North American Birds, by Elliott Coues. Fourth edition, 1890.

Manual.—A Manual of North American Birds, by Robt. Ridgway, 1887.

O. and O.—The Ornithologist and Oölogist: Birds, their nests and eggs. (A monthly publication begun in 1876, issued by several parties, published to Vol. XVIII, No. 10, October, 1893.)

Oölogist.—The Oölogist, for the student of birds, their nests and eggs. (A monthly publication begun in 1884.)

Smith. Rept.—Annual Report of the Board of Regents of the Smithsonian Institution (year, not vol.).

www.ingramcontent.com/pod-product-compliance
Lightning Source LLC
Chambersburg PA
CBHW021817190326
41518CB00007B/628